# ANALYSIS OF BIOLOGICAL MOLECULES

# ANALYSIS OF BIOLOGICAL MOLECULES

## An introduction to principles, instrumentation and techniques

**Geoffrey W.H. Potter**

*Faculty of Applied Sciences, University of the West of England, Bristol, UK*

CHAPMAN & HALL

London · Glasgow · Weinheim · New York · Tokyo · Melbourne · Madras

**Published by Chapman & Hall, 2–6 Boundary Row, London SE1 8HN, UK**

Chapman & Hall, 2–6 Boundary Row, London SE1 8HN, UK

Blackie Academic & Professional, Wester Cleddens Road, Bishopbriggs, Glasgow G64 2NZ, UK

Chapman & Hall GmbH, Pappelallee 3, 69469 Weinheim, Germany

Chapman & Hall USA, One Penn Plaza, 41st Floor, New York NY 10119, USA

Chapman & Hall Japan, ITP Japan, Kyowa Building, 3F, 2-2-1 Hirakawacho, Chiyoda-ku, Tokyo 102, Japan

Chapman & Hall Australia, Thomas Nelson Australia, 102 Dodds Street, South Melbourne, Victoria 3205, Australia

Chapman & Hall India, R. Seshadri, 32 Second Main Road, CIT East, Madras 600 035, India

First edition 1995



Typeset in 10/12pt Palatino by Best-set Typesetter Ltd., Hong Kong

Printed in Great Britain by the Alden Press, Oxford

ISBN 0 412 49050 1



A catalogue record for this book is available from the British Library

Library of Congress Catalog Card Number: 94–72005

∞ Printed on permanent acid-free text paper, manufactured in accordance with ANSI/NISO Z39.48-1992 and ANSI/NISO Z39.48-1984 (Permanence of Paper).

# Contents

Colour plates appear between pages 86 and 87.

# Preface

The majority of students in biologically related subjects will find employment in laboratories: microbiology, biochemistry, water analysis, food analysis, forensic science, environmental monitoring, quality assurance and the like. In all such laboratories increased powers of analysis have led to spectacular successes in our ability to determine the nature of the molecules that play a part in living systems and to detect them in the minutest traces. At the same time we are much more aware of the delicate nature of the quality of life and the need to monitor for deleterious agents.

Analytical science has developed specialized instruments which have become available to all scientific disciplines and there is now a bewildering array of sophisticated instruments routinely in operation in most laboratories. Such instruments use design features from physics, electronics, chemistry and other disciplines and must seem bewilderingly complex to the unfamiliar user. Nevertheless, the novice is expected to acquire skill in using the instruments very rapidly. It is probably the student from the biologically related sciences who feels most diffident about using instruments and these are the sciences which are showing the most growth.

This book presents such a student with a user-orientated approach: a combination of sufficient theory, a description of the instruments, practical advice on operating procedures and enough indication of applications to get a good start. Once this is established, the student can follow up with more specialist texts. The theory given is largely descriptive so that the student knows what lies behind the technique; necessary equations are explained rather than derived. Using routine instruments is dealt with in some detail in a manner not dependent on machine-specific instructions. An emphasis is placed on good practice and the proper handling of results, which should help the student to understand and better appreciate analytical techniques. Some indication of developments and advances are included in most chapters since the techniques are continually advancing.

It is important to state that this is **not** a textbook of practical biochemistry. Other texts present methods and procedures for dealing with various sorts of biochemical investigation and the student will be given them in pursuing any practical work. The emphasis of this book is on the tools of the trade and to what use they may be put. Again, this is not a book to describe all techniques available; it addresses only those which are routinely available and which the student will first encounter. For these reasons nuclear magnetic resonance and imaging techniques are not mentioned, while near-infrared is. Such an approach is bound to lead to some questioning of the choice made as to what to include and what to omit. The author would welcome comments and suggestions about this and, of course, on any other aspects.

The style used is intended to be approachable, starting from the assumption of little previous knowledge and meeting the reader's immediate needs. The main narrative is interspersed with numerous diagrams, as well as extra information in boxes for the more demanding reader. Margin notes and comments challenge the reader to think more about the topics rather than accept the material passively. Procedures, with 'do's and don'ts', offer useful pre-practical information so that new students can anticipate what will be expected of them in the laboratory. Students returning to a technique later in a course can use them as a summary so they need less supervision than at first.

The book is aimed at first- and second-year students on degree or BTEC HNC/D courses of any sort with some biological content, who require some familiarity with analytical methods. It provides the groundwork for such courses. It allows the tutor to use lectures and tutorials to concentrate on areas of difficulty and enables students to proceed with their practical work more efficiently. It is also suitable for anyone not on a formal course but needing to become acquainted with these techniques, such as health inspectors and environmental health officers.

Geoffrey W.H. Potter
U.W.E.

# Introduction

The remote control for my television has 36 buttons. Normally I use about six of them, which is quite sufficient for my present needs. I had to learn how to use them, of course, and when I want to use some of the more advanced features I will have to learn some more. I have looked at the instructions but have not found them helpful! I suspect many people approach laboratory instruments in a similar fashion and use them inefficiently, being intimidated by their apparent complexity and finding the instructions difficult. I hope this book will show you that you can use such machines quite easily for routine analysis and that, once you have started, you will feel confident about using them in more investigative ways.

You do need to know some of the theory as well as the practice and I have tried to give enough for an initial understanding of the methods but not so much that you give up before you start. I have deliberately been descriptive because mathematical treatments can seem very abstract to the novice. This does not mean that I dismiss such treatments as being of no use; they are essential for establishing a firm foundation for the techniques and for developing them into disciplines. But I think, at first, you are more likely to appreciate a few guiding principles than a mass of detailed relationships.

This is intended to be an essentially practical book. In a laboratory, you are expected to acquire a diverse set of skills very rapidly, and many areas of work using biological molecules need some use of analytical instruments. This is not a book which follows biochemical methods through step by step; it aims to show you what tools are available. I hope you will find that the practical procedures and theory given here will enable you to know better how analytical techniques contribute to your experimental procedures.

Don't feel you have to read the book in the order of the chapters. Use each one as you need, but you will have to do some cross-referencing because of the interdependence of the techniques. In the end you may want, or need, to know more and I have given some suggestions for further reading. There are, of course, many other titles. Almost all of them present much more theory than you will need at first. I suggest you compare similar titles and use whichever seems to make the most sense. Among the titles at the end of each chapter are some from the ACOL series (Analytical Chemistry by Open Learning). Don't be put off by the word 'chemistry': the series could well have been called 'Analysis by Open Learning'. These books are different from most since they are intended to be 'teach yourself' texts and use a very comprehensible style with questions and worked answers in the text. You could find them to be a useful addition to the material given here. Also, I make several references to some comprehensive texts which include all of the topics in this book, with far more detail either of the background chemical theory or of the components of the instruments. I have found them very useful and you might dip into them if you want to go beyond this introduction. They are Harris's *Quantitative Chemical Analysis*, Skoog and West's *Principles of Instrumental Analysis* and Braun's *Introduction to Instrumental Analysis*. Finally, two other titles which can give you some other insights into the world of analysis and instrumentation: Tyson's *Analysis: What Analytical Chemists Do* (RSC paperbacks, 1988, ISBN 0 85186 463 5) and Currell's *Instrumentation* (Wiley, 1987, ISBN 0 471 91369 3) in the ACOL series.

Details of other titles referred to are given at the ends of chapters.

**Acknowledgement**

We would like to thank Phillip Harris Scientific for the colour slides used in Plates 2, 3, 4 and 5. Plate 8 appears courtesy of Merck Ltd, UK.

# Introducing analysis 1

We talk about analysis in many contexts. We may be confronted with a personal problem and say that we need to 'analyse the situation'. Or we may have read a legal document and say that we must get a lawyer to 'analyse what it means'. From time to time we hear that astronomers have learnt more about the universe by analysing the light from the stars. Yet again, we may be concerned about the safety of a food product and suggest that it should be analysed to determine its composition.

It is this last type of analysis that we are concerned with in this book. Its meaning is similar to other uses of the word. We are taking something and separating it into its component parts, and the reason for doing this is that we believe that we will understand it better if we know what it consists of. There is, of course, a lot of truth in this but beware, lest we fall into the trap of believing that such analysis tells us everything. The whole is frequently, or perhaps always, greater than the sum of its parts.

After working through this chapter you should be able to:

- recognize various types of analysis performed in laboratories;
- list the steps in a general analytical procedure;
- state what constitutes a proper record of an analysis;
- give criteria for determining an appropriate degree of precision of data.

## 1.1 Why study analysis?

You may have asked yourself this question when you first saw the title of this book, on a bookshelf or on a reading list. I hope you haven't already 'switched off' because the topic seemed to imply boring experiments, measurements and tedious calculation. Of course there are times in analysis when a lot of detailed and routine work has to be done, but that is the case in all branches of science, and even in all branches of life. So let's start again and ask: what can analysis do that is important in your field of study?

If you are studying biology or biochemistry, you will very

quickly come across the need to examine how systems such as cells, or enzyme reactions, are affected by changes in the conditions of those systems. By 'conditions' we mean factors such as temperature, pressure, pH and, where chemical or biochemical reactions are involved, we include concentrations of reactants, products, enzymes and so on. Moreover, biological studies often require us to identify molecules that are involved in biochemical reactions, and to investigate how these reactions occur. This may sound as though you have to know a lot of chemistry as well as biology. Well, it depends on what you want your analysis to do. Many analyses can be done without detailed chemical knowledge so do not be worried about it. Nevertheless, if you are investigating at a molecular level then you will be picking up the details as you go along, just as chemists pick up biological information when they are investigating biological molecules.

For all such studies you will need to use techniques which are specially designed to aid the types of analysis needed. You will be able to interpret your experiments much better if you know how the analyses work and what they can tell you – and what they cannot, which is just as important. In fact, we might say that you shouldn't be doing an experiment unless you know something about how it works. Also, in many cases, several different methods can be used for analysing the same substance. You need to be aware of the relative advantages and disadvantages of these methods because this will help you make the best choice.

Today a major part of biological investigation needs molecular detail and the techniques described in this book have become routine tools for many workers in all sorts of biological science. For more intricate investigations the techniques have to be adapted and developed and, perhaps, specially designed. We will indicate where this is happening and refer you to other more advanced techniques described in specialist books. You will need to be familiar with the routine methods in order to get the best out of the advanced ones. Such is the pace of change in analytical science that today's advanced methods will probably soon become routine themselves.

## 1.2 What does analysis do?

Analysis is important in all branches of biology, biochemistry, biotechnology, medical science, environmental science and any other science involving biological molecules. We need to know the types of materials that we are dealing with, what molecules there are in these materials and what structures these molecules have. We also need to know how the molecules react with one another which, in turn, means knowing their concentrations, rates of reaction and how the reactions are modified if the chemical surroundings are varied. For instance, you may be fairly sure that a particular enzyme is important in the metabolism of a plant under normal conditions. But what happens when there is a drought? Is more enzyme produced or is another enzyme activated to enable the plant to survive? Could such activation be triggered by the change in concentration of a particular molecule? If so, what is the structure of this molecule?

The great achievements of molecular biology, such as the elucidation of the structure of haemoglobin and the way in which it carries oxygen or the sequence of reactions by which carbon dioxide is fixed in plants to produce sugars, have all involved a wide variety of analytical techniques. Spectrophotometry has helped to relate the subtle colour changes in haemoglobin to its oxygen-carrying ability; chromatography distinguishes between proteins that are only very slightly different; radiochemistry can unravel the complex reaction path followed by molecules in a cell and X-ray crystallography can give a three-dimensional picture of molecules containing thousands of atoms.

Because of the importance of analysis in so many branches of science, you may feel bewildered by the variety of techniques and methods and the ways in which the data are interpreted; in some cases, the amount of data is quite daunting! So, to start with, we will briefly consider how we classify analyses in order to see some similarities between various techniques, and to show that there are limitations in their applicability.

### 1.2.1 Qualitative and quantitative analysis

These are the two main kinds of analysis and you will find it useful to distinguish between them.

**Qualitative analysis** is principally concerned with determining what components are present in a sample and what substances are formed when a chemical reaction occurs. Suppose you have a solution that you have obtained by grinding some leaves with water. You may want to know if the solution contains soluble polysaccharides, sugars, acids derived from the sugars or all of these substances. You may further wish to identify some of the sugars that are present. You could also ask very similar questions about other types of substance that might be present. In answering them you would be drawing qualitative conclusions from your experiments.

**Quantitative analysis** determines how much of a particular substance is present in a sample and it may also deal with how rapidly the sample changes when subjected to particular conditions. For example, both in medicine and in sport it is important to know the level (concentration) of a drug in blood or urine

Actually: 'how much' is a vague term and we usually need to be more precise in any particular example; we may mean concentration, amount or mass and we must be clear about it!

**Figure 1.1** Official analysis of Buxton water, as shown on label.

samples, and the rate at which it may be cleared from the body. Also, we are concerned these days about the quantities of various substances in our food. Figure 1.1 shows a typical analysis of some mineral water.

There is no sharp dividing line between qualitative and quantitative analysis. Many quantitative methods are developments of qualitative tests and it is more a matter of what type of analysis is required which determines the choice of method. Some methods allow both types of analysis to be done at the same time. Chromatography is particularly useful in this way. Whatever investigation you are pursuing, there will be both qualitative and quantitative aspects to consider.

### 1.2.2 Destructive and non-destructive analysis

In many ways analysis of biological molecules has followed the tradition of chemical analysis using chemical reactions to provide the results for the analysis. This is the basis of all titrations. However, the chemical reaction also converts the substance being analysed into another substance so further analysis of that sample is likely to be much more difficult. Alternatively, the chemical conversion may give rise to a product which can be more readily analysed than the original substance (as in many colorimetric methods), but you will not be able to recover the original substance for other sorts of analysis. Such methods of analysis are known as **destructive**.

Thus, you could use a combustion method to give you the calorific value of a foodstuff, but you would need another sample if you wanted to know its glucose content. Or, if you determined the glucose content first using an enzyme reaction you would need another sample to find the calorific value because the enzyme catalyses a chemical reaction (which is destructive).

When substances are readily available, destructive analysis is acceptable. However, when a substance is available only in very limited quantity, as it frequently is in biological samples, then the loss of material must be kept to a minimum. To achieve this we must use methods of analysis which are **non-destructive**. In the main, non-destructive methods measure physical properties of substances such as the absorption of light, or their rates of movement in flowing solvents or in the presence of an applied electric field. Provided that no chemical reaction takes place during the analysis, it should be possible to use the same sample for other analyses.

Thus, if you wanted to find the lead content of roadside foliage as part of a pollution study, you would have an abundant supply of the foliage (even though the lead is present only in very small quantity) and so you could use atomic absorption spectrophoto-

**Table 1.1** Characteristics of some common analytical methods

| | *Qualitative* | *Quantitative* | *Destructive* | *Non-destructive* |
|---|---|---|---|---|
| Colour tests for sugars (e.g. Fehling's) | + | | + | |
| Infrared | + | | | + |
| Colorimetry of proteins | | + | + | |
| Flame photometry | | + | + | |
| Atomic absorption | | + | + | |
| UV spectrophotometry | | + | | + |
| Ion-specific electrodes | | + | | + |
| Chromatography | + | + | | + |
| Electrophoresis | + | + | Depends on procedure | |

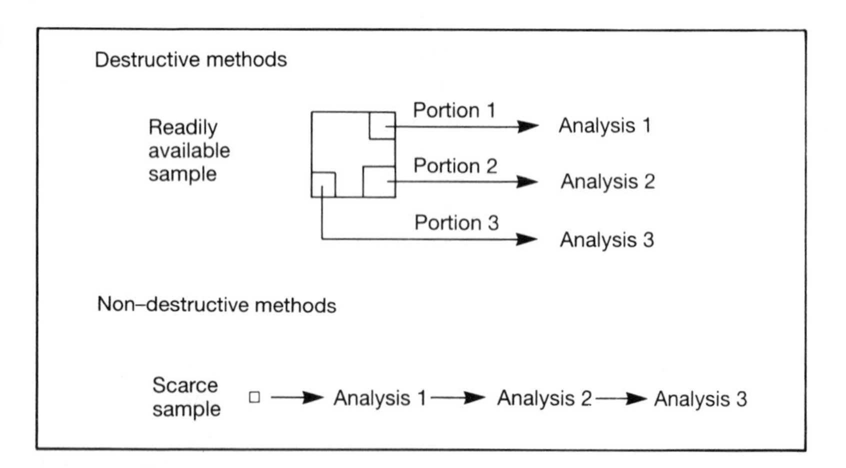


**Figure 1.2** Comparison of analytical procedures.

metry (a destructive method). On the other hand, you might be trying to establish the metabolic fate of a new pesticide using a radioactive sample. You might isolate a few micrograms of a metabolite by chromatography (non-destructive) and then attempt to establish its structure by a combination of ultraviolet (UV) and infrared spectroscopy (both non-destructive) and finally confirm it by mass spectrometry (destructive).

## 1.3 Making sure analysis is reliable

When we do any experiment we try to draw a useful conclusion from it. In analysis, our conclusion may be simply that, say, some pollutant in drinking water is at too high a level for the water to be considered safe. Or it may be that the level of protein in an animal feed is within the range declared by the manufacturer.

Look again at Figure 1.1 and try to decide what conclusions you could draw from the analyses for nitrate, iron and aluminium. In all cases we need to be certain that analytical results are reliable, and this implies that there are established criteria against which reliability can be tested.

### 1.3.1 Repeatability

The basis of all tests of scientific reliability is the idea that a properly conducted experiment can be repeated by anyone possessing suitable skills. Much is made in scientific philosophy of the principle of repeatability but, in practice, it often seems to be untrue. Students find that experiments in practical classes 'don't work', and in the research literature you may come across a phrase such as 'we were unable to reproduce the results described by . . .'.

We can still hold to the principle of repeatability, however, if we acknowledge that there are many potential variables in an experiment, and we must strive to control them as precisely as possible. Of course we need to keep records of procedures, and we must make such records as we proceed. We must also follow the analytical procedure precisely – something which is not always as easy as it sounds! If we suspect, for some reason, that results are not reliable we need to consider whether there is something inherently at fault with the method or the apparatus, or whether there was something amiss with the sample. To find out, we may have to repeat the experiment and perhaps make amendments to the recorded method.

We may think that routine analytical procedures on large numbers of samples are bound to be reliable but they must still be subject to control in order to check that we are not just repeating a mistake. We do this by checking that results from control samples are in agreement with control data. Then, if the results for our control samples agree within accepted limits with acknowledged values, we can be as sure as is humanly possible that our method is reliable.

Section 1.6 deals with how control samples are formulated and used.

## 1.4 How to deal with results

Experiments generate results, but there is no point in getting results unless you are going to use them. So a very important part of scientific work is the recording of results and correctly using them to make deductions. Let us assume that your experiment has been done properly. Then, if your deductions are to mean anything, and be reliable, you must ensure that you have accurately observed the experiment and recorded the results in a precise, unambiguous manner. The record must be in a permanent form and in such a presentation that you, or anybody else, can refer to it as necessary, even many years after the experiment was done.

### 1.4.1 Recording and reporting

When you are making notes about an experiment, or writing a full report of it, you must ensure that all the relevant results are recorded. It is very easy to leave out things that seem obvious, or things that you are sure that you will remember later on. But what is obvious to you may be obscure to someone else, and it is surprisingly difficult to remember all the details of something you did, even a few days later. You must get into the habit of always making good notes and records.

The golden rule is 'Make a proper record at the time!'

The form of a report will vary: routine analyses will probably require entries on a specially designed 'pro forma' with the description of the procedure in a separate document, but for non-routine work you will have to write up all the details in your own words. For any experiment it should be possible to compile a full report from your notes, which will include the following information:

- **title** and/or objectives;
- **name** of experimenter;
- **date**(s) on which the procedure(s) were carried out;
- **references** to related work and published information;
- **materials** used, amounts, concentrations;
- **hazard** information (this is now particularly important because of UK 'COSHH' legislation; see Section 2.2);
- **method** (in detail if it is new work but in outline or with a reference if it is repetition of previous procedures);
- **results** (raw data, readings taken, observations made);
- **graphs and calculations**;
- **conclusions** (all experiments must have at least one, even if it is only 'sample within acceptable limits' or 'inconsistent results – modify sample handling and repeat').

### 1.4.2 What sorts of results are there?

There are two types of results, qualitative and quantitative, in the same way as there are two types of analysis. If you are doing qualitative analysis your results will be mainly recorded as words, while in quantitative analysis there will be a greater emphasis on numbers.

You must make sure that words are used correctly. Remember that many words when used in scientific reports have special meanings and you must be sure that you know the proper definitions; you may not change them just to suit yourself. Also it is important to be unambiguous; for example 'clear' does not necessarily mean 'colourless' and it would be better to write 'transparent'. Another common mistake is to confuse interpretation with observation. You may well be told that when Fehling's test gives a red precipitate this is a positive result for the presence of a reducing sugar. If you see a red precipitate but just record your observation as 'positive' you are falling into this error. If someone has to consult your records, it is important that they know what you observed as well as how you interpreted the observation. They should not have to question your observation but they might question your interpretation.

Fehling's test often gives a yellow or a green colour if only a trace of reducing sugar is present. If you just recorded 'negative' or 'positive' you are not giving a proper interpretation.

Similarly, you must record quantitative values in an unambiguous manner consistent with the precision of the experiment. In all practical work there is a limit to the precision of our measurements. Mathematically we can go on putting further decimal places to a number but in practice this is either too time consuming or, more usually, we have no means of checking the validity of all the figures. In either case the extra figures may not be experimentally significant, so we will now describe what we mean by significant figures.

### 1.4.3 Significant figures

When you record a value you should write down only those figures in the measurement that you know you can obtain reproducibly. Figures that are not reproducible should not be included. By this means you indicate the precision of the experimental data.

Let us suppose that you have measured the height of a single stalk of wheat. You might report this:

- to two significant figures: 33 cm, for example, if you used a crude ruler;
- or to three significant figures: 33.2 cm if the ruler had millimetre markings.

You would even have to consider whether the stalk were stretching a little while you were measuring it! This takes us onto the realm of experimental error, which we discuss further in Section 1.5.

It is unlikely that you could expect to measure the stalk any more precisely than this since you could not be sure that you could estimate fractions of a millimetre reproducibly.

### 1.4.4 Decimal places and significant figures

The concept of significant figures can be confusing when zeros appear in a number. Suppose you had weighed some caffeine extracted from a coffee sample and obtained the following display on the balance:

0.0052 g.

This value has two significant figures. The leading zeros show only the magnitude of the number and are not regarded as significant in the above meaning of the term. Also, if the balance does not give you any further figures (to the right) you are not justified in writing 0.00520 g since that would imply that you knew for sure that the weight was not 0.00521 g or 0.00519 g. You could use scientific notation and write the above value as $5.2 \times 10^{-3}$ g, which clearly shows that the precision is two significant figures.

Similarly, with values greater than 10, such as 45 600, you should regard the two zeros as indicating only the magnitude, so this number has three significant figures.

Beware! In the UK and USA this will probably be written as 45,600 but in continental Europe the comma indicates the decimal point! To avoid the confusion, the latest convention is to indicate only the decimal point (by a comma or a full stop) but you can leave a space between groups of three digits to make the number more readable.

Of course it could have four or even five, and if it were important to be unambiguous about it you should use scientific notation: $4.56 \times 10^4$ for three significant figures, $4.560 \times 10^4$ for four and so on.

Although scientific notation can be clearer than the decimal point form, it is rather cumbersome for everyday use and much of the time you will find it easier to use the appropriate prefixes, such as milli and kilo, with the units. Thus, it might be best if you wrote the above weight as 5.2 mg, which again shows clearly two significant figures. For the same reason as above, 5.20 mg (three significant figures) is not appropriate for the measurement.

**Box 1.1 Calculators are a nuisance!**

Of course calculators save us a lot of tedious calculation and when used properly will gives us a correct answer, but:

- They show far more significant figures (eight, ten or even twelve!) than can be justified by the precision of almost all experimental work, and it is wrong (as well as a waste of time) to write them all down slavishly. Most analytical work is done to a precision that requires three or four significant figures (see Section 1.5 on experimental uncertainty for more about this).
- They generally give a display which is a garbled version of scientific notation. You will need to get used to a display such as 1.23 – 02 which actually means $1.23 \times 10^{-2}$ and you will need to be sure how to enter such a value (often by means of a button designated 'exp'). It is easy to be wrong by a factor of ten: entering 1.23 × 10 'exp' −02 = gives 0.123 since the 'exp' includes the . . . ×10. Try it and see!

### 1.4.5 Units

Most analytical measurements determine quantities. We will define a quantity as the product of a number and a unit of measurement:

$$\text{quantity} = \text{number} \times \text{unit}$$

For example, for an average adult male

$$\text{height} = 1.8 \times \text{metre}$$

and you could write this in symbols:

$$h = 1.8\,\mathrm{m}$$

The system of units which gives the greatest consistency for scientific work in general is the *Système International d'Unités* (SI units). All sorts of other systems of units have been used and a few of their units are still referred to: calorie (cal), Ångström unit (Å) and curie (Ci), for example.

Find out what these units are if you do not know already.

**Table 1.2** Common SI units

Make sure you get the case (upper or lower) of the letters correct – it is important

**The seven fundamental units**

| *Quantity* | *Unit* | *Symbol* | *Quantity* | *Unit* | *Symbol* |
|---|---|---|---|---|---|
| Mass | kilogram | kg | Temperature | kelvin | K |
| Length | metre | m | Amount of substance | mole | mol |
| Time | second | s | Luminous intensity | candela | cd |
| Electric current | ampere | A | | | |

**Derived units** may have their own symbol but are ultimately expressible in terms of the fundamental units

| *Quantity* | *Unit* | *Symbol* | *Definition* |
|---|---|---|---|
| Area | metre squared | | $\mathrm{m^2}$ |
| Volume | metre cubed* | | $\mathrm{m^3}$ |
| Velocity | metres per second | | $\mathrm{m\,s^{-1}}$ |
| Frequency | hertz | Hz | $\mathrm{s^{-1}}$ |
| Energy | joule | J | $\mathrm{kg\,m^2\,s^{-2}}$ |
| Force | newton | N | $\mathrm{J\,m^{-1}}$ or $\mathrm{kg\,m\,s^{-2}}$ |
| Pressure | pascal | Pa | $\mathrm{N\,m^{-2}}$ or $\mathrm{kg\,m^{-1}\,s^{-2}}$ |
| Power | watt | W | $\mathrm{J\,s^{-1}}$ or $\mathrm{kg\,m^2\,s^{-3}}$ |
| Potential difference | volt | V | $\mathrm{J\,A^{-1}\,s^{-1}}$ |
| Amount of electricity (charge) | coulomb | C | $\mathrm{A\,s}$ |
| Activity (of a radioactive substance) | becquerel | Bq | $\mathrm{s^{-1}}$ |

* note that the litre (l or in America L) is now defined as a cubic decimetre ($\mathrm{dm^3}$)

**Prefixes** – multiples and sub-multiples of units are designated by the addition of an extra letter. Note that the kilogram is anomalously named!

| *Multiple* | *Prefix* | *Symbol* | *Sub-multiple* | *Prefix* | *Symbol* |
|---|---|---|---|---|---|
| $10^{18}$ | exa | E | $10^{-18}$ | atto | a |
| $10^{15}$ | peta | P | $10^{-15}$ | femto | f |
| $10^{12}$ | tera | T | $10^{-12}$ | pico | p |
| $10^{9}$ | giga | G | $10^{-9}$ | nano | n |
| $10^{6}$ | mega | M | $10^{-6}$ | micro | $\mu$ |
| $10^{3}$ | kilo | k | $10^{-3}$ | milli | m |
| | | | $10^{-2}$ | centi[a] | c[a] |
| | | | $10^{-1}$ | deci[a] | d[a] |

[a] Only in cm and dm.

On the whole it is better to use only SI units, but if you have a particular reason for using any others in a report you should also give the SI equivalent or a conversion factor. Table 1.2 summarizes the units you are likely to have to use in analysis and there are some conversion factors for non-SI units in Appendix 1.

### 1.4.6 Tables

When you have made a number of measurements it is often convenient to present the data in the form of a table. You should record only the numbers in the body of the table; the quantities and their units should be in the headings. This avoids a lot of unnecessary repetition which can be quite distracting. You can also include values that you have calculated from data in other parts of the table. When you do this you should show one example of the calculation separately from the table, but there is no need to show them all. The results of the calculations which you put in the table can easily be checked by referring to the example.

**Box 1.2 Word processors and spreadsheets**

Word processors and spreadsheets are extremely useful computer programs for the scientist but they do have some limitations that you should be aware of. Here are a few points.

- **Word processors** are good for producing reports of experiments when there is a lot of description and discussion. You can vary the type and the layout of the paragraphs to make the report easy to follow and you can used **bold-face type** if there is something in the middle of a paragraph that you want to draw attention to. You should learn to use all of the common facilities, but do give yourself time to practise before trying to use them on that urgent report that must be handed in tomorrow! When printing out make sure that the printer gives the same page breaks as those on screen and that columns stay in line. Use double spacing of lines to make the text easier to read. Tables are not easy to generate with word processors unless the package has a special facility for this. Superscripts, subscripts, chemical and mathematical formulae also need special facilities. When you are reasonably familiar with the common facilities you should learn how to import material from other files and applications such as spreadsheets and drawing packages.
- A **spreadsheet** is like an array of calculators. Data are entered along with the necessary formulae and many interrelated calculations can be done very rapidly. Changes and recalculation are easily done. With suitable formulae you can mimic chemical and biological changes so that you can explore experiments on screen (mathematical modelling). Parts or whole spreadsheets can be incorporated into word-processed documents along with bar charts, histograms and the like, which can greatly enhance your reports. Note, however, that graphs containing lines of best fit may not be easily obtainable from simpler spreadsheets and you may need to use a curve-fitting package.

Whatever type of package you are using you have to be careful if you are trying to transfer it to another computer. Unless it is using exactly the same version of the package you may find that there are difficulties such as wrong formatting.

**Table 1.3** Specimen table of results

**Evolution of gas during fermentation**

| *Clock time (min:s)* | *Elapsed time (s)* | *Volume of gas ($cm^3$)* | *Rate of gas production ($cm^3 s^{-1}$)* |
|---|---|---|---|
| 5:01 | 301 | 10.0 | 0.0332 |
| 10:10 | 610 | 20.1 | 0.0330 |
| 15:04 | 904 | 29.9 | 0.0331 |
| 20:25 | 1225 | 40.7 | 0.0332 |
| 25:00 | 1500 | 50.1 | 0.0333 |
| 30:05 | 1805 | 59.9 | 0.0332 |

**Typical calculations:**
Convert clock time to SI units: 5 min 1 s = 5 × 60 + 1 = 301 s
Rate of production of gas = volume/time = $10.0\,cm^3/301\,s = 0.0332\,cm^3\,s^{-1}$

*Note*: Units should be put in the headings within parentheses as shown; they should not be repeated all the way down the column. (You may find other conventions, so take care that you are interpreting them correctly.)
Very large and very small numbers give clarity problems in tables. Scientific notation does not help resolve this and your best method is to choose the units with care, for example mmol instead of mol.
The number of significant figures quoted represents the approximate precision.

If you were testing whether gas from a fermentation was being produced at a constant rate you would be noting the time and measuring the volume of the gas. A table of your results could look like that shown in Table 1.3.

### 1.4.7 Graphs

There are several points that you need to take care of when you draw graphs.

1. **Which axis is which?** You use the horizontal $x$-axis for the quantity that you have been changing during the experiment (the independent variable), and the vertical $y$-axis for the quantity that you have been measuring as a result of the changes (the dependent variable). Notice that when we talk about drawing a graph of, say, volume against time, this means the volume is on the $y$-axis and the time is on the $x$-axis.
2. **The labelling of the axes** can be done in the same way as the column headings in tables. Remember that it is important to show the units.
3. **The range of values** on each axis should be chosen carefully. You need to be sure whether it is important to include the origin (0,0) or not. You can check the theory of the method to help you decide. If you are drawing a graph with logarithmic values it is unlikely that (0,0) will be an important point since a log value of 0 corresponds to an original value of 1, which is much less likely to be a value of particular significance.

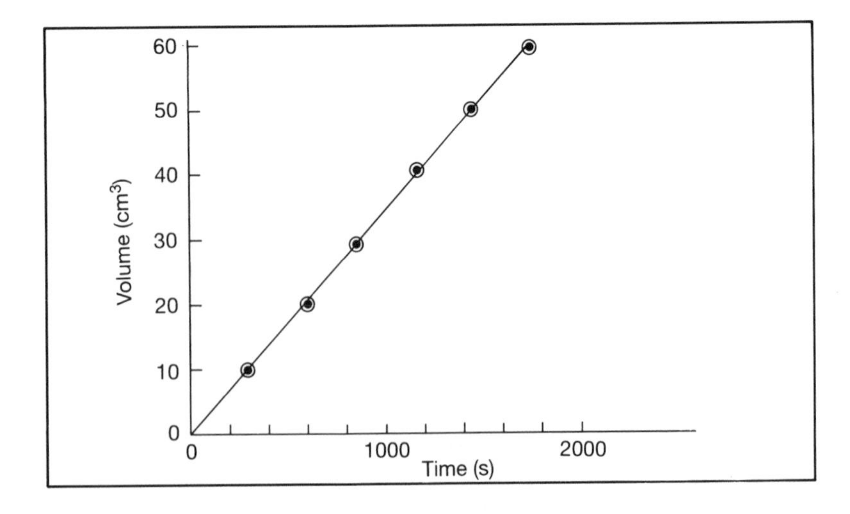


**Figure 1.3** Volume of gas evolved during a fermentation.

4. When you are **fitting a line to points**, allow for experimental uncertainty and draw your line so that it has an even scatter of points each side. You need to know whether you are expecting the line to go through the origin and whether it should be straight or not. If it is a curve, draw the smoothest one you can with as few twists and turns as possible. Since a straight line gives greater precision, many methods will mathematically process the data in order to produce a straight line (linearization). For the sorts of graph we use in analysis we are always looking for the line of best fit. It is never appropriate to 'join the dots', although you will find this done in other sorts of experiments.
5. Always put a **descriptive title** to the graph so that the reader knows what the graph actually shows.

Thus, for the gas volume results given above, you might draw the graph shown in Figure 1.3: note that there is no evidence of curvature in the points so the straight line of best fit is drawn.

## 1.5 Experimental uncertainty and error

The theory of uncertainty in measurement is summed up in Heisenberg's uncertainty principle. He was a German physicist (1901–1976) who contributed to the development of quantum mechanics. The principle states that we can never precisely know both the position and energy of a particle. The more precise we are about one, the less precise we have to be about the other!

However hard we try, we can never make absolutely precise measurements. This is not just human fallibility but a fact of nature: there is always some uncertainty in observation and measurement. Even with the most refined apparatus and the greatest skill there is a natural limit to the precision of a measurement.

In all the experimental work we are considering in this book, this absolute uncertainty is very small compared with the uncertainty arising from human fallibility and imperfections in apparatus. This combination of factors is the **experimental uncertainty**. It is often called **experimental error** but this is a bit unfair since we should be able to do experiments without actual errors of procedure, but there is still some limit of precision beyond which we cannot make the results reproducible.

This means that we cannot really use the term 'exact value'; all we can do experimentally is to determine a value with as little uncertainty as possible. From here on we will use the term **true value** for the value which we would obtain if experimental uncertainty were reduced to the absolute minimum. Our best estimate of the true value is the **mean**. This is the statistical term for what is commonly called the average. It is found by adding up all the experimental values and dividing by the number of them. The more measurements that contribute to the mean, the better the estimate.

### 1.5.1 Random and systematic uncertainty

We divide the sources of experimental uncertainty into two sorts.

1. ***Random uncertainty*** (or **random error**) is the result of unattributable fluctuations within the system as a whole that may increase or decrease any measurement from the true value. Large deviations are less common and small deviations more common. The effect of random deviations can be reduced by repeating the measurements a number of times and finding their mean.
2. ***Systematic uncertainty*** (or **systematic error**) affects all the measurements in the same way and cannot be reduced by finding the mean of replicate values. It is symptomatic of a bias in the measurements (perhaps caused by an error in weighing, or an error in setting the zero of an instrument, for example). It may not be apparent until the results are compared with those from a different method.

Statistical techniques are used to analyse the ways in which values are spread about their mean by using various mathematical expressions to describe these distributions (see Box 1.3). It also provides us with methods for assessing the significance of experimental values. The mathematics of statistics can seem very abstract and rather daunting but we can give a description of what the two main statistical parameters represent, and these days, fortunately, we can use calculators or computers to do the donkey work.

When you quote an experimental value you should indicate the amount of uncertainty you feel is present. One way of doing this is to quote the **range** of values that were used to determine the

## Box 1.3 What some statistical terms mean

When a large number of measurements of a quantity is made, statistics gives several methods of deriving a representative value. The most widely used is the **mean** (or **average**), obtained by adding together all the values and dividing by the number of values used. Your calculator may do this automatically for you. Find out if it will! Other representative values are the **median** (the value which has equal numbers of other values above and below it) and the **mode** (which is the most frequently obtained value).

The spread of values about a mean can be indicated by a **range** (the highest and lowest values obtained) but the most useful statistical measure is the **standard deviation**, which is the average square root of the squares of the deviations from the mean! Once again, find out how to do this automatically on your calculator. You will probably find that there are two keys: one labelled $\sigma_n$ (or just $\sigma$) and the other $\sigma_{(n-1)}$ (or $s$). Since you will only be using a restricted set of data you should use the second of these.

The distribution about the mean of the values obtained can be pictured using a frequency diagram. The shape of the diagrams may correspond to one of a number of mathematical expressions. The most frequently encountered ones are:

- the **Poisson distribution**, which applies to measurements that have discrete values and cannot be less than zero;
- the **normal distribution**, which applies to measurements that can take any value (continuous variable).

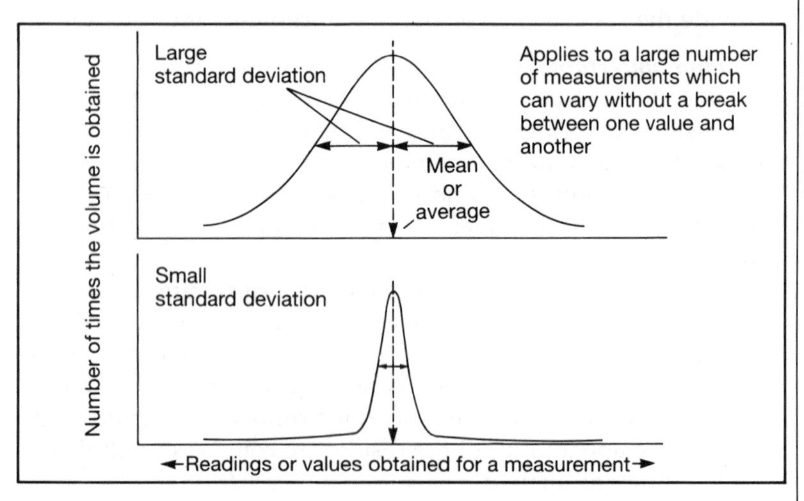


**Figure 1.4** The normal distribution.

In the normal distribution 68% of all the values lie within the region up to one standard deviation each side of the mean.

You can find a full introduction to these topics in T.A. Watt's *Introductory Statistics for Biologists*, mentioned in the 'Further reading' section at the end of the chapter.

mean. This gives some idea of how far your values spread each side of the mean by indicating the highest and lowest measurements made, but it can be misleading if you have one or two values which are widely different from the rest. Statisticians use the **standard deviation** to give a better measure of the spread of the values. It gives more emphasis to the more consistent values and is less affected by extremes.

For much analytical work described in this book the standard deviation is much less than the mean and our simplified treatment of the statistics is sufficient. In other sorts of biological investigation, however, such as the correlation of human height and weight, or the growth rate, say, of wheat when compared with the boron content of the soil, there is a much greater degree of uncertainty and the standard deviation may be approximately equal to the mean. For these cases you have to use other, more appropriate, statistical methods.

Find out how to obtain the mean and the appropriate standard deviation (see Box 1.3) on your calculator or computer. You do not have to understand the mathematics behind the calculations. Just get used to what the values describe.

### 1.5.2 Accuracy and precision

At this stage we need to be aware of a particular usage of two words which have a special meaning in analytical experiments, and to relate them to the statistical terms we have just discussed.

1. ***Accuracy*** means how close an experimental value is to the true value. Note that in fundamental science, such as determining the velocity of light, estimates of true values must be consistent with a tested theory. In analytical science, where you might be determining the level of vitamin C in a foodstuff, there is no theory to predict the value. You make your best estimate of the true value by comparison with a standard that has been produced under strict control.
2. ***Precision*** is the term used to indicate how consistent repeated measurements are.

Try not to confuse the two terms. In everyday usage they are often regarded as synonymous. Can you see that you can have high precision but poor accuracy? Table 1.4 gives some figures to illustrate this. Analysts are always trying to improve both accuracy and precision.

For a given experimental method, there are statistical methods of improving the accuracy. The simplest and most familiar method is to do repeated measurements and determine the mean (take an average). The mean is more likely to be closer to the true value. Of course it takes time to do more measurements so it may not be an option when results of analyses are required rapidly. Or, you may not have enough material for many repeated analyses, so there has to be a compromise between accuracy and the practical constraints.

The precision of a set of values is the experimental equivalent of the standard deviation. A smaller standard deviation indicates a greater precision. In contrast to the mean, we cannot expect the precision to be improved by a larger number of repeated measurements. Improvements can come only through increased skill and

**Table 1.4** Accuracy and precision

Suppose four laboratories were asked to determine the volume of 1 mol of an ideal gas under conditions of standard temperature and pressure. The true value is 22.4 $dm^3$. If each laboratory produced six values as follows, we could judge their accuracy and precision thus:

| *Values* | | | |
|---|---|---|---|
| Laboratory 1 | | | |
| 22.3 22.5 22.7 | Mean | 22.4 | Good accuracy |
| 22.2 22.3 22.4 | St. dev. | 0.18 | Good precision |
| Laboratory 2 | | | |
| 20.4 22.2 24.4 | Mean | 22.4 | Good accuracy |
| 22.6 23.2 21.6 | St. dev. | 1.37 | Poor precision |
| Laboratory 3 | | | |
| 23.3 23.5 23.7 | Mean | 23.4 | Poor accuracy |
| 23.2 23.3 23.4 | St. dev. | 0.18 | Good precision |
| Laboratory 4 | | | |
| 21.4 23.2 25.4 | Mean | 23.4 | Poor accuracy |
| 23.6 24.2 22.6 | St. dev. | 1.37 | Poor precision |

more refined methods. Lack of precision in analysis tends to imply lack of reliability in a method, so it is important to be able to specify the precision you attain. You can best do this by quoting the standard deviation but, as with the mean, you have to have replicate values for this. Otherwise you can just quote the mean to an appropriate number of significant figures. In biological analysis you will find that three significant figures corresponds to the approximate precision of your analysis. This is explained a bit more in the section below on proportional uncertainty.

Statistically, a value quoted as $1.23 \pm 0.04$ means that we are 68% certain that the true value is within the range 0.04 above or below 1.23 (see Box 1.3). Just quoting 1.23 would indicate that the true value probably lies between 1.225 and 1.235 but we would not put a percentage figure to the probability.

**Box 1.4 Discardable values?**

Occasionally you will find that you have obtained a value which is so different from what is expected that you suspect some gross error has occurred. Statistically, there is a small chance that such a value could have occurred randomly but you will probably think it more likely that it was caused by some error in your procedure and you may feel entitled to discard it.

Some analysts are guided by whether a result 'looks wrong' but this is very subjective. They might decide that the correlation of all other points on a calibration graph is so good that one which does not correlate well must be in error. Others use mathematical tests for determining whether discarding might be justified, such as discarding a value because it is more than three times the standard deviation away from the mean.

Statisticians do not agree that this is entirely legitimate since the usual methods should accommodate such values, but they do allow that if you can show that there was definitely an error in the procedure for that point you can discard it. With experience, rules of thumb give at least a warning that some of the data may be unreliable and you should do some further analyses before deciding how to interpret the results.

### 1.5.3 Proportional uncertainty (or proportional error)

It is important to relate the amount of uncertainty to the value itself. What might seem to be a very small uncertainty could be a large proportion of the value itself. In such a case the uncertainty would be of distinct significance to the conclusions you might draw from the result.

Suppose you had obtained a mean value for the concentration of sodium ions in tap water of 0.0010 mol $dm^{-3}$, and had calculated that the standard deviation of the determinations was 0.0004 mol $dm^{-3}$. This seems to be a very small spread of values but it is actually 40% of the mean. So you would be saying that there is a 68% probability that the true value lies between 0.0006 and 0.0014 mol $dm^{-3}$ (see Box 1.3) and hence a 32% probability that it lies beyond these limits! Not really such good precision as you might at first have thought. You are less likely to mistake the importance of the uncertainty if you use the sub-multiple mmol $dm^{-3}$ instead of the original unit. You would then write $1.0 \pm 0.4$ mmol $dm^{-3}$ and the proportionality is more obvious.

If you are relying only on significant figures to indicate the precision of your results, remember that a possible variation of one in the last digit is implied by the convention. So if you quote 1.23 (three significant figures) as a value then a variation of one in the last figure is just under 1% of the 1.23, so the uncertainty is about 1%. You might find that this is the sort of precision you obtain at first in analytical techniques, but if you were working on routine analyses you would expect to get a greater precision and would probably quote four significant figures, 1.234, which would imply about 0.1% precision.

$(0.01/1.23) \times 100\% = 0.813\%$

and

$(0.001/1.234) \times 100\% = 0.081\%$

You could try finding the percentage precision for the values of the standard deviations given in Table 1.4.

### 1.5.4 Combinations of experimental uncertainties

In many procedures you have to make several measurements in order to be able to calculate the value you require. The proportional uncertainty in each of these may be different. In many experiments you have to weigh samples, dissolve them in measured volumes of solvent and then use instruments to make further measurements. You can generally weigh samples with high precision (a sample of weight around 1 g weighed to the nearest 0.1 mg is being weighed to 0.01% precision), while burettes can at best be read to about 0.5% precision. This is discussed more fully in Section 2.5.

When such measurements are combined in a calculation, then the **overall precision can be taken to be roughly that of the least precise measurement**. Thus a combination of the above weighing and burette readings would have a precision of approximately 0.5%. Suppose you had found that the volume of ethanoic acid required to just neutralize 1.000 g (0.1% precision) of sodium carbonate was 20.0 $cm^3$ (0.5% precision). You could readily calculate that 100 g of carbonate would probably require 2000 $cm^3$ (2 $dm^3$), but you would have to recognize that the true value could be between 1990 (2000 − 0.5%) and 2020 $cm^3$ (2000 + 0.5%).

Precise figures for uncertainty combinations can be determined mathematically, but are not really needed except in very refined work.

## 1.6 Calibration and control

There are three main components in an analysis:

1. setting up and calibrating the procedure;
2. using the method for test samples;
3. using control samples to check that the method is reliable.

### 1.6.1 Setting up and calibrating

In setting up an analytical method, you may have to follow a fully described procedure, or adapt one so that it suits your samples, or you may even have to devise a method of your own. Whichever of these is the case, you can see that it is important to keep a full record, from the start, of what materials you use and what you actually do.

We described what constitutes a proper experimental record in Section 1.4.

You should not be too surprised if the method does not work first time. Faults can be present in even the simplest experiment. Once it is running as required, the next step is calibration and for this you need a blank and standards.

You might consider how you would know whether an experiment is running as required or not!

### 1.6.2 Blanks

An important feature of analytical experiments is that you must establish an appropriate zero reading for your method. Even if there is none of the substance being analysed in a sample, an unadjusted instrument may still give a non-zero reading. Because of this, you start with such a sample (a **blank**) and adjust the instrument to make it read zero, rather than use unadjusted readings. As a general rule, blanks should contain everything that is in the test sample except the substance for which the analysis is being set up.

For example, if you are using a colorimeter to measure protein concentrations by the so-called biuret reaction, you should make up a blank containing the reagent and salts and other substances which are normally present, but no protein. In this case, the blank is sky-blue and so does absorb some visible light, but when you have chosen an appropriate wavelength you set the colorimeter to read zero, since it is the change from blue to purple that you need to measure, not the blue itself.

### 1.6.3 Standards

Calibration is the process by which you relate the experimental readings to the actual quantities of the substances you are trying to measure. If the analysis is qualitative the calibration may be more properly called correlation but the essence of the process is the same. Calibration implies that you have samples of known composition from which you can establish the relationship between the experimental results and the compositions. These samples are known as **standards**. Once you have done the calibration you can take readings or make observations on the test samples whose composition is to be determined. It is important that your standards are made up in a way which closely resembles the test samples because the presence of other substances in a solution can alter the results obtained, in which case your analyses might be erroneous and would certainly be less reliable.

### 1.6.4 Calibration range

Of course, you do not know the actual values of the test samples you are going to analyse, but it is helpful to have some idea of the range of values within which they might fall so that you can prepare a set of standards to cover the expected range. There are times when a method is applicable only to a more limited range of values. In such cases, you usually dilute the test samples so that their values come within the calibrated range. Occasionally you have to make a sample more concentrated in order to obtain reliable values.

### 1.6.5 Determining the concentrations or composition of test samples

You will have done some sort of sample preparation (see Section 2.7) on the original material to produce the test samples. You then use the same measuring procedure as for the standards, and you may do replicate measurements to reduce the experimental uncertainty. You can then use an appropriate formula using the test results and those from the standards, or a calibration graph, to

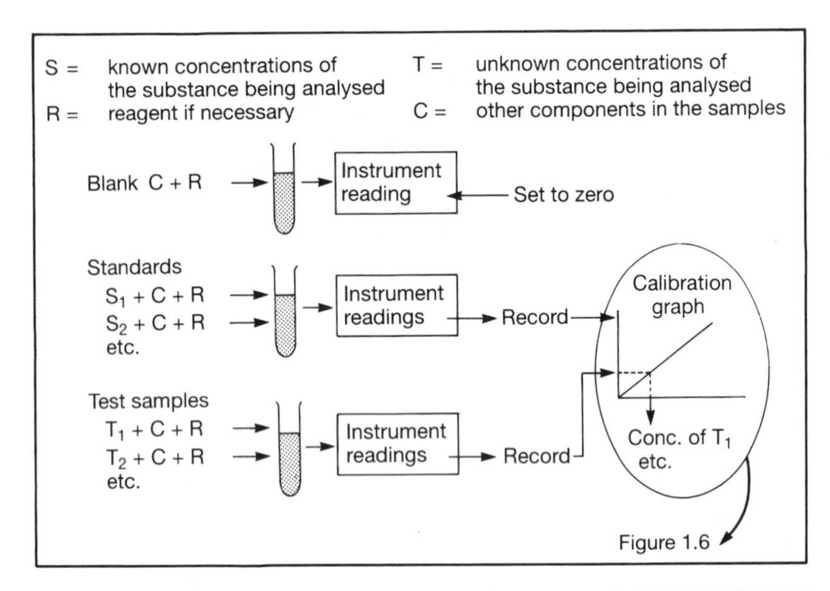


**Figure 1.5** Procedure for analysis.

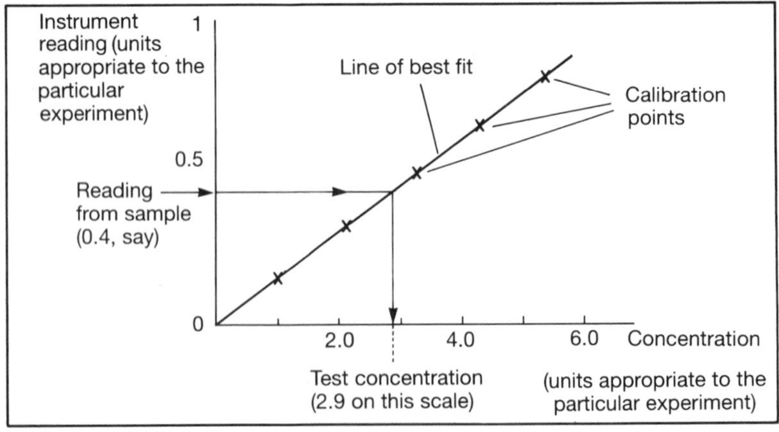


**Figure 1.6** Calibration graph.

determine the concentration or composition of the tests samples. Figure 1.5 gives an indication how this is done and you can find examples in later sections on specific methods such as colorimetry (Section 3.1 and Figure 3.5), flame photometry (Section 3.7.1 and Figure 3.21) and gas chromatography (Section 5.3 and Figure 5.6b).

It is important to realize that values outside a calibrated range may not be reliable because you have not demonstrated that the calibration or correlation still holds. But once you have obtained readings for diluted samples as mentioned above, you can apply the dilution factor to determine the concentration of the original sample. Suppose you had had to dilute your sample 10-fold in order to obtain an instrument reading which was within the calibrated range. You would then determine the concentration in the diluted sample, probably by using a graph such as that shown in Figure 1.6, and multiply by 10 to get the original concentration. In that figure you can see that the diluted sample has a concentration of 2.9 units, so the original would have been 29.0 units.

In flame photometry, for example, calibrations are most reliable at very low concentrations (see Section 3.7). Samples such as blood, serum or milk contain sodium levels which are too high for reliable direct measurement. Dilutions of 50- to 500-fold may have to be made before the levels in the original samples can be measured reliably.

### 1.6.6 Standard additions

Standard additions are an alternative to simple calibration. If you are analysing complex materials, you may find it difficult to formulate standards that closely mimic the test samples. You may not know all the components in the samples. In such a case, you can use a procedure known as **standard additions**. Unlike simple calibration, you start by measuring a test sample. Then you add a known amount of the pure substance to it and remeasure the sample. Usually, you do one or two further additions and take the corresponding measurements. Finally, by drawing a graph as shown in Figure 1.7 you can determine the original concentration of the sample. The advantage of the method is that any effects due to other components in the test sample are automatically allowed for in the results. A disadvantage is that each test sample has to be measured several times. The method only works reliably if the graph is a straight line because it requires extrapolation. Curves do not give reliable extrapolations.

This type of calibration is used widely in atomic absorption spectrophotometry (Section 3.7).

Extrapolation is the projection of a graph beyond the measured points. Can you see why curves are difficult to extrapolate? If you allow that there is experimental uncertainty in all the points then you can draw many lines that look like curves of best fit, but go in different directions beyond the points.

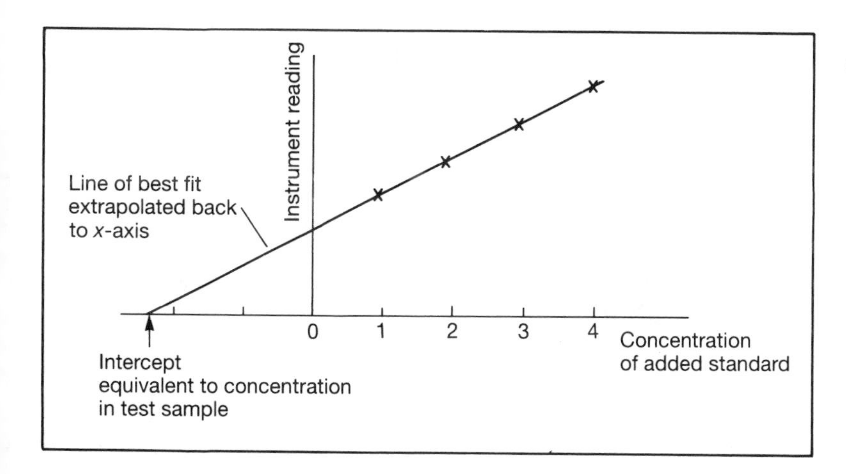


**Figure 1.7** A typical graph for the standard additions procedure.

### 1.6.7 Controls

You use controls to keep a check on the reliability of your analyses. They are samples containing a known concentration of the substance you are analysing but are not part of the calibration. They are made up to be as similar to test samples as possible and you analyse them in exactly the same way as the test samples. You can use either simple calibration or standard additions procedure. Their results should fall within a range of acceptability specified for that control. If they do not, your method may have to be recalibrated.

For the greatest reliability, it is important to ensure there is no operator bias. In laboratories regularly performing analyses it is usually arranged that the operator does not know which samples are test and which are controls, and the controls occur randomly throughout the procedure. Details of which sample is which must be kept by someone who is not involved in the actual analysis.

## Further reading

Harris, D.C. (1987) *Quantitative Chemical Analysis*, W.H. Freeman, ISBN 0 716 71817 0.
Chapters 1, 3 and 4 are useful supplementary material for the topics in this chapter.

For a full treatment of the statistics see:

Caulcutt, R. and Boddy, R. (1983) *Statistics for Analytical Chemists*, Chapman & Hall, ISBN 0 412 23730 X.
Gives a good feel for the statistics needed for analysis whatever sort of molecules you are dealing with.

Watt, T.A. (1993) *Introductory Statistics for Biologists*, Chapman & Hall, ISBN 0 412 47150 7.
Don't forget that the statistics needed for biological experiments are somewhat different from those needed in analysis, so be prepared to pick and choose what is relevant here.

The ACOL 'teach yourself' style books include the following titles.

McCormick, D. and Roach, A. (1987) *Measurement, Statistics and Computing*, Wiley, ISBN 0 471 91367 7.

Woodget, B.W. and Cooper, D. (1987) *Samples and Standards*, Wiley, ISBN 0 471 91290 5.

# 2 Working in laboratories

What is special about working in laboratories? You may be wondering why we should even bother to ask this question. After all, many occupations require special premises or facilities. Also, you may have already done some laboratory work and think that the laboratory was nothing more than work-benches with gas taps and high stools. On the other hand, there are undoubtedly laboratories where very specialized work is done, often with some sort of prohibition of entry to unauthorized persons.

In popular fiction, films and television, laboratory workers are often portrayed as peculiar people wearing white coats, doing strange things with weird apparatus. Indeed, this tradition goes back at least a hundred years: there is an element of madness or obsession in all of H.G. Wells' fictional scientists for example.

But there are all sorts of laboratories and laboratory workers. There should be no particular mystique about the work, nor any stereotype. The same applies to almost all occupations. That does not mean to say that laboratories and laboratory work cannot have a special attraction or excitement. It really depends on how interested you become in the work.

After working through this chapter you should:

- recognize the importance of safe working;
- be able to list good working practices;
- know how to weigh samples, measure volumes and perform other routine manipulations;
- be aware of procedures for suitably preparing analytical samples.

## 2.1 What is special about analytical laboratory work?

Within each of the main disciplines – biology, chemistry and physics – there are general-purpose laboratories, but also there are many types of laboratory designed to cater for some sort of special purpose. Nevertheless, there are areas of overlap, and you will find some instruments and apparatus in, say, both chemistry and biology laboratories. This is particularly true where analysis is done, and it is often arranged that an analytical laboratory serves several disciplines, especially if it houses complex instruments such as are needed for chromatography and spectroscopy. Also, a division may be made into laboratories for preparation of samples ('wet labs') and those for the measurement of the samples ('dry labs') so that the sensitive apparatus is not subjected to contamination.

Analysis these days is just as much concerned with biological samples as with those of non-biological origin. In many cases in biological work you will be using analytical methods developed by chemists, and you will need to know something about them in order to use them properly. There is no longer any clear distinction between biology, biochemistry, biological chemistry and chemistry.

You may think that analysis implies repetitive routines producing nothing but numbers! On the other hand, in modern analytical laboratories there are machines with keyboards and screens, automatic samplers and print-outs which cost many thousands of pounds. A large proportion of biological analysis uses such machines, but what do they do? One of the aims of this book is to give you some insight into what some of these machines are used for and how they work, which should be interesting in itself. However, we also have to acknowledge that there are times when, even with these machines, there is much repetition and routine operation, but if you know how the analysis works and what the results mean in biological terms, if you are looking for ways of improving a technique or seeking to correlate your results with other data, then analysis can be interesting and challenging. However, if you are going to produce good work you have got to obey the ground rules.

## 2.2 Working safely in laboratories

The materials you handle in laboratories often have particular hazards associated with them, but procedures have been devised to minimize the risks in using them. You will encounter apparatus that you are not familiar with. This means that a laboratory is a special place and has rules for safety which you must observe, for your own sake as well for others'. A laboratory must be treated with respect; it is not a place where you can do as you please or have a bit of fun!

This is a serious point! Do take proper note of it.

Nowadays, we are all very conscious of safety, and rightly so.

In many countries it is a legal requirement that employees work in a safe manner, and make sure that their fellow workers are able to do so as well. Of course, there are times when the rules and procedures seem to be more trouble than they are worth, and there always seems to be someone in the laboratory who does not bother. Your job is to bother! Get used to best practice straightaway to save having to learn a better way later.

You might compare this with seat-belt legislation.

General laboratory rules include many common-sense things which you may feel you can ignore: don't! Look at the list given in Table 2.1 and ask yourself why each rule has been included. Also think about what might happen if you deliberately ignore a rule. The likelihood of accidents happening may be very small, but the consequences can be severe, so the risk is not negligible.

Specialized laboratories, such as those where radioactive

**Table 2.1** Laboratory safety rules

**General**

No eating, drinking, chewing or smoking in the laboratory

Wear laboratory coats at all times, and safety glasses and other protective clothing or devices when necessary. Make sure you are used to working with such protection; if necessary, practise with non-hazardous materials and equipment first.

Know where the fire exits are, as well as the fire-fighting equipment.

Know the laboratory rules for dealing with fires.

Know the procedure for dealing with other accidents.

Carry large bottles in a proper carrier. Dispense only the minimum quantity required.

Check that there are no flames in the laboratory when using flammable liquids.

Check that no flammable materials are in the area when you light a Bunsen burner.

Never pipette by mouth, and take particular care when inserting a pipette into a pipette filler.[a]

Take particular care when attaching a rubber tube to any glass apparatus.[a]

Use fume-hoods for operations whenever necessary, even though it may take a little more time to set up the operation.

Dispose of waste solvents and solutions in an approved manner (you will need to find out what this is in any particular laboratory).

**Specialist laboratories**

**Radiochemical laboratories**: rules additional to those above may include the following

Shielding of an appropriate nature must surround both open and sealed $\gamma$-sources.

All work with open sources must be carried out in trays lined with a suitable absorbent material.

**Microbiological laboratories**: rules additional to the general ones may include the following

Protective clothing must be changed at least once a week.

No one may enter a virology laboratory without the permission of the person in charge.

[a] Broken glass is the commonest of laboratory accidents. Always wet the glass when attaching any fitting to it, and do not overtighten clamps on it.

material is handled, require extra rules and particular care must be given to minimizing any radiation dose acquired by the laboratory workers, as well as the prevention of contamination of the laboratory and its surroundings. Similarly, in microbiology laboratories, where dangerous pathogens may be handled, special rules will be applicable. Some examples of such extra rules are given in Table 2.1.

If you are working in the UK there is legislation which must be observed. The Health and Safety at Work Act (1974) requires us all (whether in a laboratory or not) to adopt safe procedures and work in such a way as not to endanger ourselves or others. You cannot 'opt out' of this – for example, you cannot decide you are not going to wear safety glasses because you think they look ridiculous. If you are not prepared to wear the correct safety gear you will not be allowed to do the experiment.

### 2.2.1 The COSHH regulations

The Control of Substances Hazardous to Health (COSHH) Act (1988) requires an assessment to be made of the risks from using any substances with known hazards in any procedure. For routine procedures, this will have been done by someone with responsibility for that procedure but, if it is a new procedure, you may have to do the risk assessment yourself before starting (with guidance, of course, from people who have already done such an assessment, and with reference to books and databases containing the hazard data).

Before you start an experiment you must:

- note warnings given in instruction sheets;
- look for warnings on labels and wall charts;
- ask supervisory staff;
- consult hazard data sheets where appropriate.

Of course, all substances can be harmful in some way or other. For the purposes of laboratory work the term 'harmful' includes substances which are:

- toxic, narcotic/soporific/addictive, pathogenic;
- can cause tissue damage or irritation (especially to the eyes/ respiritory tract/skin).

It is important that you know the levels (concentrations) at which such effects might occur, and also the total doses. Before you begin an experiment you should know, therefore, whether each substance is harmful in microgram, milligram, gram or multigram quantities, and you should take into account whether you will be handling the neat substance, small quantities of concentrated solutions, or large quantities of dilute solutions. Do not

**Table 2.2** COSHH data for an experiment introducing the use of a chloride electrode

| *Materials* | *Hazards* |
|---|---|
| **Standards** | |
| Potassium chloride (0.10 mol $dm^{-3}$) | May be emetic in quantity |
| **Test samples** | |
| Unknown potassium chloride solution (sample X) | May be emetic in quantity |
| Tap water | Non-toxic |
| Mineral water (commercial) | Non-toxic |
| Serum substitute | Non-toxic |

**Control measures**: normal laboratory practice

**Risk assessment**: insignificant in this procedure

forget that dilute solutions dry out to give concentrated solutions, and thus become more hazardous. Thus you must mop up spillages in an approriate manner, as soon as possible.

For the purposes of risk assessment, it will be assumed that you are following a normal laboratory code of practice anyway, so only the extra hazards will be brought to your attention.

An examples of a risk assessment is given in Table 2.2.

## 2.3 Working methodically

When you are learning a new procedure, you expect to be given a set of instructions for each experiment. A good set will indicate precisely what is to be done, and there is a temptation to presume that you can start straight away. In fact, this is a very inefficient way of working since you way well find that, in the middle of a procedure, you need solutions, say, that are not readily available which you should have prepared beforehand. It is far better to spend a few minutes first making sure you know what the experiment is about, what apparatus is needed (and whether it is available!) and what materials and samples need preparing; also what results need to be recorded and what the best format for this is.

When you are more experienced and are attempting experiments that are not routine, you will have to work out a procedure for yourself (probably with much discussion with supervisors and colleagues). It will be based on previous experience and on published work. Once again you will find it best to plan the work carefully even though it seems, at first, that you are not 'getting on'.

You could devise a check-list, such as:

- aims
- information I will need
- things I will need
- plan of experimental work.

The result might look something like this:

Aims
To determine the concentration of magnesium in tap water samples by atomic absorption

Information I will need
wavelength of measurement, gas mixture for flame
preparation of stock solution, concentration range of standards
likely sample concentration range, treatment of samples

Things I will need
lead nitrate (analytical purity)
5% nitric acid (analytical purity)
plenty of high-purity deionized water

Plan of experimental work
prepare 5% nitric acid from concentrated acid
calculate weight of lead nitrate needed for stock solution
prepare lead nitrate solution
make dilutions for standards
prepare samples
set up atomic absorption instrument
set blank
measure standards
measure samples

## 2.4 Working analytically

When you apply the above principles to analysis you may need to take the following points into account.

**For qualitative analyses:**

- Are you looking for known or unknown components in a sample?
- Are you looking for major or trace components?
- Have you a sufficiently sensitive method?
- Can you identify the components of interest, or just classify them according to type?
- Will the method discriminate sufficiently between all the possibilities?
- Ought you to use further methods to provide supplementary or supporting evidence?

**For quantitative analyses:**

- What are the limits to the reliability of the method?
- What sort of treatment does the sample require?
- Will other components in the sample interfere?
- Have you established a suitable regime of blanks, standards and controls?

**Concerning choice of method:**

- Are there alternative methods?
- How well do they compare in ease of use, sensitivity, detection limit, selectivity?

**Reporting of results:**

- What results are to be recorded?
- To what precision should they be recorded?
- What is the overall precision of the method?
- Are graphs required?
- How are they best presented?

**Drawing conclusions:**

- Have the correct statistical methods been applied to the results?
- Are the results consistent with known or expected values, or ranges or qualitative interpretations?
- If not, has the sample been found to be significantly different from what was expected?
- Is there a fault in the method, in the instrument or the apparatus?

## 2.5 Basic analytical procedures

### 2.5.1 Weighing

This is the most precise operation you are likely to use, and these days it is one of the simplest. Digital display top-pan balances with a tare facility are rapidly taking the place of the older (although just as precise) 'dial-a-weight' balances. In either case all you have to do is:

- **ensure you have a balance of the appropriate range** (for example 160 g max., or larger) **and of the appropriate precision** (low precision is ±0.01 g; high precision is ±0.0001 g, which is ±0.1 mg);
- **close any doors in the balance cover** to avoid draughts (particularly necessary with high-precision balances);
- **zero the display** (with a suitable container on the pan);
- **add the sample to the container** and close the doors again;

You may well come across the instruction 'weigh accurately about 2 g of . . .' This is not a contradiction in terms. It means that any weight near 2 g will be acceptable, as long as you know precisely what the weight is.

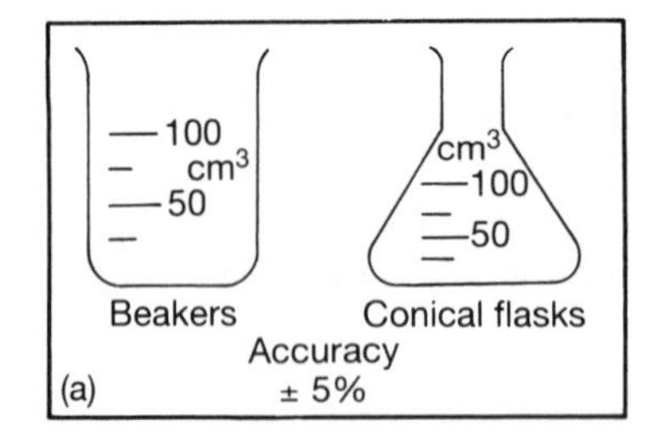


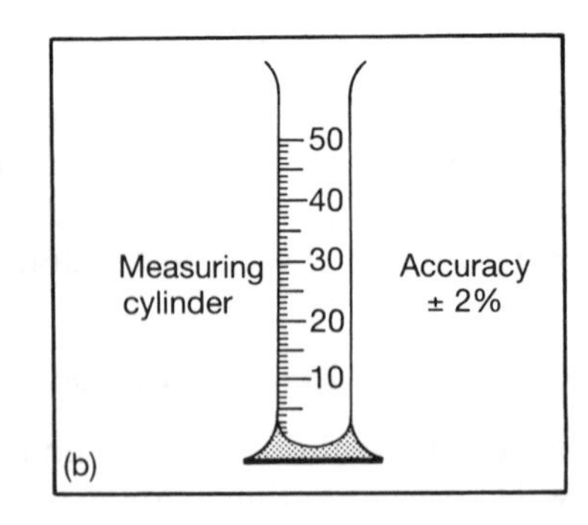


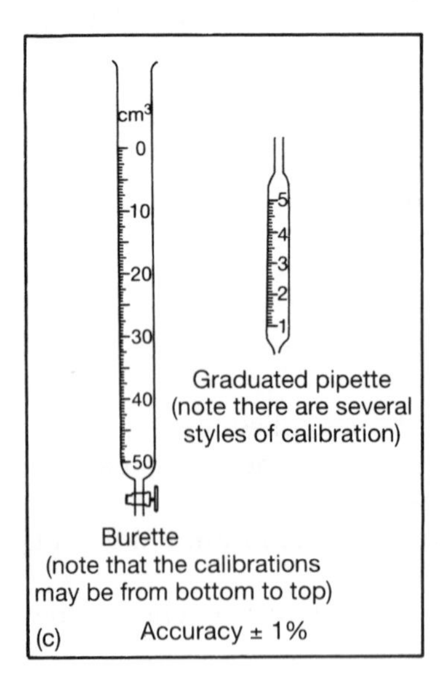


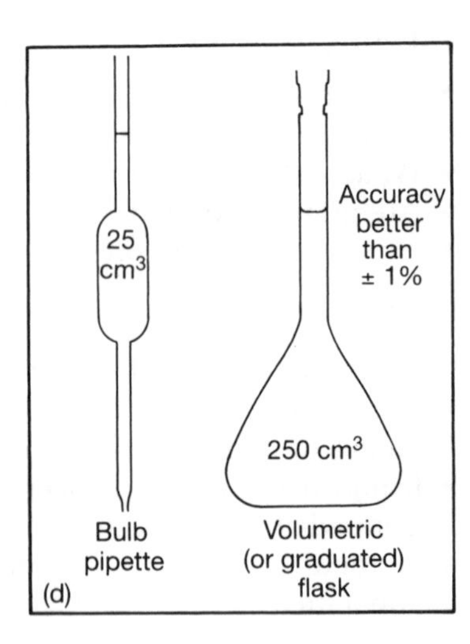


**Figure 2.1** Volumetric glassware: (a) of rough accuracy; (b) of moderate accuracy; (c) of good accuracy; and (d) of highest accuracy.

- **read the weight** (and record in a suitable manner; not on your coat sleeve or the back of your hand!).

### 2.5.2 Measuring volumes

This is one of the least precise operations that you do in a laboratory, so it is important that you use the most appropriate method for the precision required (Figure 2.1). Note that graduated pipettes are being replaced by a variety of piston action pipettes where the volume may be dialled up on the plunger. These give superior precision but are very expensive.

### 2.5.3 Preparation of solutions

Many biological and chemical processes are dependent on the concentration of one or more of the components in the system and quantitative analytical procedures are designed to determine one or more of them. However, in order to follow the analytical

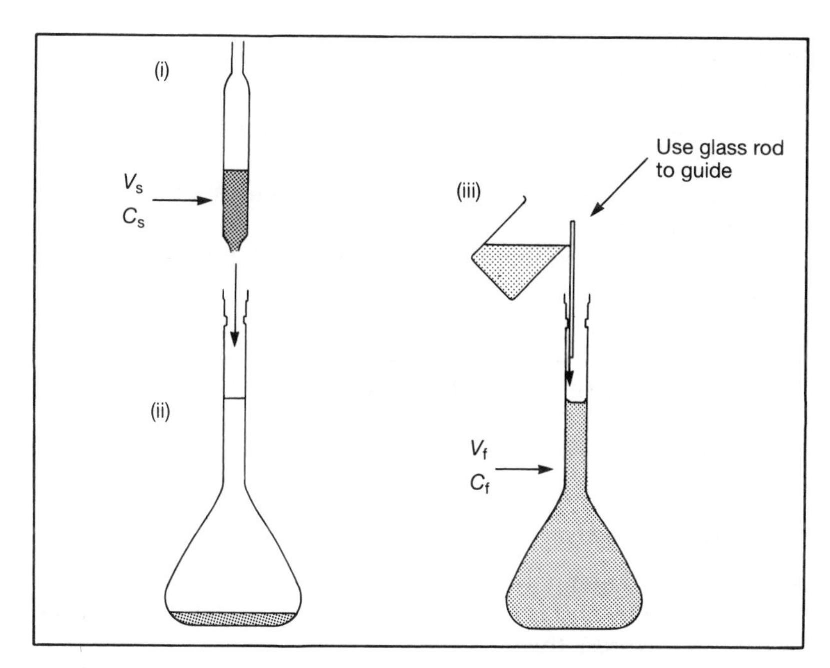


**Figure 2.2** Making a dilution (see below and 2.6.2): (i) measure volume of starting solution; (ii) transfer starting solution to volumetric flask; (iii) add solvent from beaker to make up to graduation mark. Add the last few drops from a 'squeezy bottle' so that the bottom of the meniscus lies on the graduation line.

procedure you must be able to prepare and manipulate solutions to produce known concentrations. You will have to know the weights and volumes you will need for a particular experiment and we will discuss how to do the calculations in Section 2.6. For now, we will illustrate the procedures.

*Preparation from solid materials*

If you are starting from a solid, you will have to weigh out a portion of solid and dissolve it in a known volume of solvent. The following method is best.

1. **Weigh the solid** in a suitable container (such as a small plastic 'boat') using an appropriate degree of precision.
2. **Transfer the solid to a small beaker** and add some solvent. Add rinsings from the boat to ensure complete transfer. Stir the mixture until all the solid is dissolved.
3. **Transfer the resulting solution to a volumetric flask** using a glass rod as shown in Figure 2.2. Transfer the rinsings as well.
4. **Make the volume up to the calibration mark** with more solvent.
5. **Mix the solution thoroughly** by repeated inversion of the flask.

*Preparation from liquid materials*

This is usually known as **making a dilution**. For experiments involving **precise** analytical work you must use the following method. The reason for its importance is given in Section 2.6 where we calculate concentrations. Refer to Figure 2.2.

1. **Accurately pipette a volume of stock solution into a volumetric flask.**
2. **Make the solution up to the calibration mark with the appropriate solvent.**
3. **Mix the solution thoroughly** by repeated inversion of the flask.

## 2.6 Calculating concentrations and dilutions

Concentrations are usually defined in two related ways.

**Non-SI units**
The use of **$l^{-1}$** and **$ml^{-1}$** instead of **$dm^{-3}$** and **$cm^{-3}$** is still widespread but is discouraged by SI rules.
**ppm**, 'parts per million', is also widely used. It corresponds to mg $dm^{-3}$.
**ppb** is 'parts per billion' and becoming popular as detection levels get lower. You need to remember that it is the American usage of 'billion' ($10^9$) so it corresponds to $\mu g$ $dm^{-3}$.
Also, **percentage concentration** is grams per 100 $cm^3$; often abbrieviated to % (w/v).
Non-SI units should be regarded as obsolete, even if they seem to be useful. They can be ambiguous. It is better to convert to SI and get used to a consistent set of units.

*When masses (weights) are important*

The concentration in terms of mass (weight) is known as the **mass concentration ($c$)** and is calculated as **mass ($m$)** divided by **volume ($V$)** and we express this mathematically as:

$$c = \frac{m}{V}$$

Commonly used units are **grams per litre**, written as **g $dm^{-3}$**. Note that this is the same as **milligrams per millilitre**, written **mg $cm^{-3}$**.

For dilute solutions you may use **milligrams per litre (mg $dm^{-3}$)**, which is the same as **micrograms per millilitre ($\mu$g $cm^{-3}$)**.

For very dilute solutions you may use **micrograms per litre ($\mu$g $dm^{-3}$)**, which is the same as **nanograms per millilitre (ng $cm^{-3}$)**.

Make sure that you recognize clearly which units you are using.

*When the amount (the number of moles) is important*

A very widely used unit is **molar concentration ($C_m$)**, which is defined as the number of **moles ($n$) per litre** and has units **mol $dm^{-3}$**. Remember that this is the same as **millimoles per millilitre (mmol $cm^{-3}$)** and as with mass concentrations it can be helpful to think in these terms. Putting this in mathematical terms we can say:

$$C_m = \frac{n}{V}$$

In biological work you will frequently use **millimolar concentration** which is the number of millimoles per litre (**mmol $dm^{-3}$**) and is the same as **micromoles per millilitre ($\mu$mol $cm^{-3}$)**.

In practical terms, you are more likely to weigh out a sample and then have to calculate the molar concentration, so you can use the relation:

$$\text{number of moles} = \frac{\text{mass in grams}}{\text{relative molecular mass}}$$

or, in symbols:

$$n = \frac{m}{M_R}$$

with corresponding expressions in terms of millimoles and milligrams. This leads to the combined expression:

$$C_m = \frac{m}{M_R \times V}$$

### 2.6.1 Stock solutions

You are most likely to need these formulae when making up **stock solutions** which are used for storage and dispensing of common reagents. They are usually more concentrated than is needed for the experiments, which reduces the storage space needed. Another advantage is that they are often less susceptible to deterioration than dilute solutions. If you are starting with a solid, take into account the volume of solution you may need, and use whichever of the above formulae is appropriate to calculate the weight of substance required. If you are starting with a liquid, it is easier to measure its volume. You will then need to know its density so that you can calculate its mass and then proceed as for a solid.

### 2.6.2 Making dilutions

*Starting from solutions of known concentration*

Having made up a stock solution, or been provided with one, you usually have to prepare other concentrations from it. Refer again to Figure 2.2 for an illustration of the procedure. For any dilution operation, you can always calculate what you need to know by using the following formula. As long as you can put values to three of the variables you can calculate the fourth.

$$C_f V_f = C_s V_s$$

where $C_s$ = concentration and $V_s$ = volume taken of **starting** solution, and $C_f$ = concentration and $V_f$ = final volume of **diluted** solution.

Remember that the units of $C_f$ and $C_s$ must be the same, as must those of $V_f$ and $V_s$.

Do not confuse this equation with a similar one for chemical reactions which is often written in terms of $M_1V_1$ etc. and deals with reacting amounts.

For example, you may have to make up $100\,cm^3$ ($V_f$) of a sodium chloride solution at a concentration of $0.05\,mol\,dm^{-3}$ ($C_f$) from a stock solution of concentration $1\,mol\,dm^{-3}$ ($C_s$). What volume of the stock solution ($V_s$) must you use?

You can rearrange the equation to read:

$$V_s = \frac{C_f \times V_f}{C_s}$$

and hence calculate that

$$V_s = \frac{0.05\,mol\,dm^{-3} \times 100\,cm^3}{1\,mol\,dm^{-3}} = 5\,cm^3$$

This is not the same as adding $95\,cm^3$ water since there may be some volume change on mixing (see below).

so you should take $5\,cm^3$ of the stock and make it up to $100\,cm^3$ with distilled or deionized water and thoroughly mix the resulting solution.

Alternatively, you may be told to dilute $10\,cm^3$ ($V_s$) of a stock solution of, say, copper sulphate to $250\,cm^3$ ($V_f$). If the concentration of the stock ($C_s$) were $50\,g\,dm^{-3}$, what concentration would you obtain ($C_f$)?

A similar calculation to the previous one gives:

$$C_f = \frac{C_s \times V_s}{V_f} = \frac{50\,g\,dm^{-3} \times 10\,cm^3}{250\,cm^3} = 2\,g\,dm^{-3}$$

*Using a dilution factor*

Alternatively, you may be asked to make a 10-fold dilution of a stock solution, or to dilute it 1 to 10. This does not mean that you must start with $1\,cm^3$ and finish with $10\,cm^3$. It simply means that the final volume must be 10 times the starting volume. The **ratio of $C_f$ to $C_s$ is known as the dilution factor.**

For example, you could pipette

| | | | |
|---|---|---|---|
| | $1\,cm^3$ into a $10\,cm^3$ flask | and add solvent to give a total of | $10\,cm^3$ |
| or | $5\,cm^3$ into a $50\,cm^3$ flask | | $50\,cm^3$ |
| or | $10\,cm^3$ into a $100\,cm^3$ flask | | $100\,cm^3$ |

If you are making up $50\,cm^3$, say, of a such a solution you should **not** record

volume of stock solution: $5\,cm^3$
volume of water: $45\,cm^3$

for two reasons:

1. The volume of water is not measured as such.
2. The mixing may produce a change in the total volume.

It is proper to record 'volume of stock $5\,cm^3$ made up to $50\,cm^3$ with water' (or whatever solvent was used).

or any other combination which gives you the volume you require and uses pipettes and flasks of convenient sizes.

Notice that in all these calculations the volumes that are important are the starting volume ($V_s$) and the final (or total) volume ($V_f$). You may well expect this final volume to be the sum of the individual volumes used, but this is not always the case! It is important to realize that **volumes can change during mixing,** especially if concentrated and dilute solutions are mixed, or if solvents such as alcohol are present. Thus we should stress that the 10-fold dilution (1:10) mentioned above is not the same as '1 plus 9' since there may be volume changes because of the mixing.

However, in biological work, where the precision of the experiment may not be as great as for analytical work, it is often inconvenient to use the 'making up to the mark' procedure, especially when preparing a variety of reaction mixtures. In these cases, the '1 plus 9' type procedure **is** used and the final or total volume **is** assumed to be the sum of the components. Since such reaction components are usually dilute aqueous solutions the procedure is regarded as a justifiable approximation.

### 2.6.3 Sets of standard concentrations or sets of dilutions

You often need to make up a range of concentrations of a standard so that you can obtain a good calibration for your method. There are two particular ways of making up such sets of concentrations.

*Linear range of concentrations*

This has equally spaced values and is used when the effect to be measured is directly proportional to the concentration, such as in a colorimetric assay. If you needed, say, $100\,cm^3$ of each of a set of glucose concentrations, 0.2, 0.4, 0.6, 0.8, $1.0\,mmol\,dm^{-3}$, and had a stock solution of concentration $10\,mmol\,dm^{-3}$ you could achieve this by diluting the following volumes of the stock solution to $100\,cm^3$ using an appropriate solvent:

You can check that you will get these concentrations by using the formula $C_fV_f = C_sV_s$.

$$2,\ 4,\ 6,\ 8,\ 10\,cm^3$$

*Logarithmic range of concentrations*

This is known as serial dilution and produces a series of concentrations in which each is related to the next by the same factor. It is done by starting with the stock solution, making a first dilution and using it for the second which is in turn used for the third, and so on. The procedure is used when the effect to be measured is logarithmically related to the concentration as, for example, in the use of ion-selective electrodes or in some microbiological assays.

So if the dilution factor were 10 and you started with a stock solution of an antibiotic of concentration $100\,mg\,cm^{-3}$ you would obtain the following concentrations:

Again, check that you will get these concentrations by using the formula $C_fV_f = C_sV_s$.

$$100,\ 10,\ 1,\ 0.1,\ 0.01\,mg\,cm^{-3}$$

Note that, although you may make dilutions of solutions, the important quantity to record is the **concentration** that results from the dilution. You often have to plot a graph of some measured quantity against concentration in order to calibrate an instrument or a method.

## 2.7 Sample preparation

The various types of analytical technique differ widely in their requirement for sample preparation. Some can deal with quite crude samples that contain many other components. In such cases there must be no interference from these other substances. In contrast, where there could be distinct interference the sample may have to be purified to some extent or the substance of interest be extracted from the crude sample. In these cases it is important that the procedure is monitored to check the efficiency of the extraction.

### 2.7.1 Types of crude sample

*Waters*

These may be the easiest to deal with since they will be fairly dilute. You may have to filter or centrifuge them to remove suspended solid material. If you need to analyse for traces of, say, pesticides or oil residues it may be necessary to concentrate the sample by evaporating the water or by sampling the vapour above the water.

*Body fluids*

Urine, serum and other fluids are less complex than cellular materials. Proteins, amino acids, sugars, inorganic ions and so on may be analysed directly or may require the protein, say, to be precipitated. Alternatively, dialysis through a membrane can hold back large molecules, allowing smaller ones to pass into the solution to be analysed.

*Cellular materials*

Almost invariably the complexity of cellular materials precludes their direct analysis. In particular, the membranes in a cell may localize some substances and prevent their true level being established. It is necessary in these cases to break up the cellular structure and liberate the substances of interest. Of course, it is important that the procedure does not harm the molecules you want to analyse. Small molecules are not damaged by physical processes such as grinding or homogenizing, but large ones – proteins and particularly DNA – may be.

*Foodstuffs*

Cellular integrity is not important in foodstuffs so they may be treated in the same way as for cellular materials but with less emphasis on preserving delicate molecules. Some foodstuffs are less complex and sometimes quite highly purified – butter and sugar for example – so their analysis may require little or no pretreatment.

### 2.7.2 Techniques

Where cellular materials have tough membranes or cell walls you can **grind the material** with sand in a mortar. If you add a solvent the soluble substances will be extracted. The extraction of chlorophyll from leaves into methanol is a good example of this. An alternative is to use a heavy-duty kitchen liquidizer which will chop up the material with rapidly rotating sharp blades.

**Homogenization** is a more gentle form of grinding or liquidizing. The most widely used type is a thick-walled test-tube of very uniform diameter which has a close-fitting plunger. The membranous components of the cells are broken up as they are squeezed between the plunger and the wall of the tube. The bulk of the macromolecular structures in the cells are not destroyed although DNA and proteins may be denatured.

Extracts and original samples are often too concentrated to be measured directly. They have to be diluted to a known extent with an appropriate solvent. We have already discussed **dilution** procedures in the preceding section.

Some samples, however, are too dilute for direct measurement and have to be made more **concentrated**. Various evaporation techniques can be used, or, in some cases, volatile vapours can be allowed to collect in the space above the sample and then analysed (head-space analysis).

You may find that large molecules in your sample, such as proteins, polysaccharides or DNA, make the solution cloudy or bind (attach strongly) certain ions. These sorts of effects prevent you from obtaining a true analysis, but you can avoid the difficulty by **precipitating** the macromolecules with reagents such as tungstic acid or solvents such as ethanol.

An alternative method of removal of large molecules is **dialysis**. A dialysis membrane holds back the large molecules but allows the smaller ones to pass into solvent in which the small molecules can be analysed. You may, however, need to analyse the large molecules but not have the small ones present. In such a case you retain the solution which has not diffused through the membrane.

This is sometimes called 'de-salting'.

**Filtration** can be done through a variety of materials: paper, glass wool, muslin and so on. The retained material may be sticky or gel-like and may clog up the filter. The filter aid Celite (a diatomaceous earth) may help to prevent such clogging. Some filter systems are 'cross-flow', which also helps avoid clogging.

Another alternative is **centrifugation**. There are no fine pores which may become clogged and the intense centrifugal field exaggerates small differences in density, so that different materials settle at different rates with coarser or denser particles collecting at the bottom of the tube most rapidly. Small molecule do not settle out at all, but large molecules will if very high speeds are used. The separations may be aided by doing the centrifugation in a specially prepared solution whose density varies from top to bottom of the tube (density gradient centrifugation).

Breaking up the cellular structure by grinding or homogenizing may not be sufficient to release all the small molecules from cellular material. In such cases you have to use acidic or alkaline **digestion**. The organic material is destroyed and so the small molecules are completely liberated.

Which procedures you use depends, of course, on the nature of the crude sample and the analyses that you wish to carry out. Here are two examples.

1. **Determination of magnesium in spinach leaves:** You could dry the leaves and weigh them, cut them into small pieces, digest the pieces by heating them with 50% nitric acid until all the carbon has been oxidized, dilute the resulting solution with deionized water, filter and then use atomic absorption spectroscopy.
2. **Determination of the sugars in honey:** You could weigh a small portion of honey and dissolve it in enough deionized water to give an approximately 1% solution. A further 10-fold dilution with 50% acetonitrile followed by membrane filtration would give a solution that could be analysed by high-performance liquid chromatography.

## Further reading

Harris, D.C. (1987) *Quantitative Chemical Analysis*, 2nd edn, W.H. Freeman, ISBN 0 716 71817 0.
Fundamental laboratory techniques are described in Chapter 2.

Robyt, J.F. and White, B.J. (1987) *Biochemical Techniques in Theory and Practice*, Brookes/Cole, ISBN 0 534 07944 X.
Chapter 2 of this text also gives similar details.

# Spectroscopic techniques

# 3

We are all familiar with colour and its use to enhance our lives aesthetically. We also use it to impart information without resorting to words; in appropriate circumstances red means 'danger' while green may mean 'go'. Another use, perhaps subconscious, is that we can gauge 'how much' by colour, e.g. a bright or a dark colour may mean 'a lot of', while a pale colour may mean 'weak'.

Here, then, are the beginnings of colorimetry, an analytical technique involving colour measurements which tell us something about the nature or identity of a sample, and also how much it contains.

When we make measurements that go beyond the range of the visible spectrum, into the ultraviolet (UV) or the infrared (IR), or even beyond these, we can gain even more information about the sample. The principles remain the same but the subject becomes spectroscopy.

After working through this chapter you should be able to:

- describe the nature of electromagnetic radiation in terms of wavelength and energy;
- account for the interaction of such radiation with matter and the analytical uses of these interactions;
- outline the construction of various types of spectrophotometers;
- understand the principles of operation and calibration of the instruments;
- apply procedures for making quantitative analyses;
- interpret qualitative information with respect to molecular structure.

## 3.1 Colours and colorimetry

### 3.1.1 The nature of light

What do we know about light from everyday observation? It is given out by very hot objects such as the sun, or the filament of a light bulb. It can travel through materials that we call transparent, but not through materials that are opaque. It can also travel through materials such as coloured glass, but then its colour may become modified and it becomes diminished in intensity. When a beam of white light passes through a prism it becomes spread out (dispersed) into the colours of the visible spectrum. You see the same effect in a rainbow since the sunlight passing through raindrops at certain angles is dispersed in a similar fashion. You can also see a spectrum when white light is reflected from a surface which has many fine lines close together, such as a gramophone record or a compact disc. The scientific name for such a reflector specially made to do this is a 'diffraction grating'.

These properties of light are best explained by the wave theory of light (see Box 3.1). The waves are in the electric and magnetic fields that are everywhere in the universe. As with all wave motions, the velocity (the speed of propagation) of the wave and its frequency (the number of oscillations per second) are important and characteristic. However, mainly for historical reasons, when we deal with light in analytical work we describe the wave in terms of its wavelength. Different colours have different wavelengths and white light contains all the visible wavelengths.

**Box 3.1 Characteristics of electromagnetic waves**

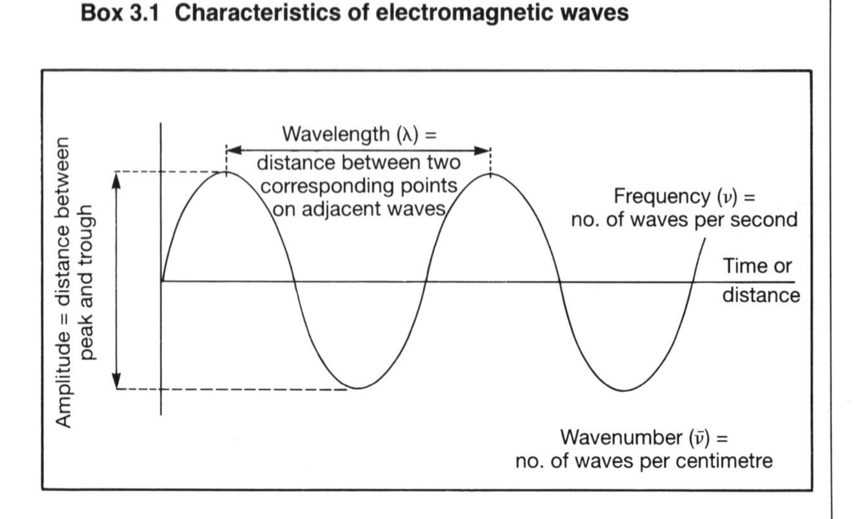


**Figure 3.1** Characteristics of electromagnetic wave motion.

The relationship between the wavelength, the velocity and the frequency for any wave motion is:

$$\text{wavelength (m)} = \frac{\text{velocity (m s}^{-1}\text{)}}{\text{frequency (s}^{-1}\text{)}}$$

or in symbols

$$\lambda = \frac{c}{v}$$

The unit of frequency is referred to as hertz (Hz) but its dimension is $s^{-1}$. For visible light and all other types of electromagnetic radiation (see Section 3.2 for more details) it is the frequency of the radiation which is its fundamental characteristic. Its velocity depends on the material through which the radiation is passing, which in turn affects the wavelength. To avoid confusion in analysis, the wavelength we refer to is that which the radiation would have in a vacuum. In this case its velocity is the 'speed of light,' which has a value of $3.00 \times 10^8$ m $s^{-1}$ (to three significant figures). The wavenumber is a figure which is used widely in IR spectroscopy but not otherwise in analysis. Here are some typical figures for the UV, visible and the IR regions of the spectrum:

| *Region* | *Wavelength (nm)* | *Frequency (Hz)* | *Wavenumber ($cm^{-1}$)* |
|---|---|---|---|
| UV | 250 | $1.2 \times 10^{15}$ | 40 000 |
| Visible (yellow) | 590 | $5.17 \times 10^{14}$ | 17 241 |
| IR | 5000 (5 μm) | $6.0 \times 10^{13}$ | 2 000 |

The dispersion by a prism or a diffraction grating happens because each wavelength is refracted or diffracted at a slightly different angle. It is important to note that the material of the prism or grating is not itself coloured; it does not add or subtract anything from the light. The visible spectrum is traditionally thought of as containing seven main colours with all the shades in between. Some colours, brown for example, do not seem to be in the spectrum but this is because of the way our eyes respond to colours. In addition to this, we are usually looking at colours which contain ranges of wavelengths, rather than single wavelengths. It is, nevertheless, quite convenient for descriptive purposes to associate the seven colours of the rainbow with their approximate wavelengths as shown in Plate 1.

When white light shines onto coloured material, the pigment in the material either reflects or transmits only the wavelengths that we see. What happens to the wavelengths that are not reflected or transmitted? These wavelengths are absorbed by the pigment and the colour we see is white light minus what is absorbed. Thus grass looks green because the chlorophyll pigments absorb a range of wavelengths in the blue region of the spectrum and another range in the red region, which leaves a range of mainly green wavelengths. The precise range and intensity of the wavelengths give the different shades of green.

Can you deduce what colours are absorbed by other pigments such as the yellow colour in butter (carotene), or the red colour of haemoglobin?

### 3.1.2 The absorption spectrum of a sample

In spectroscopic terms we speak of chlorophyll having two **absorption bands**: one in the range approximately 400–500 nm and the other 600–700 nm. The upper part of Figure 3.2 gives a picture of the absorption bands of chlorophyll, but it is also important in analysis to have a numerical measure of the amount of absorption at each wavelength. This can be done using a colorimeter, which measures the intensity of the light transmitted through the sample in relation to the intensity of the light entering it. This ratio may be expressed as the **percentage transmittance** ($T\%$); however, it is generally more useful to use another scale known as **absorbance** ($A$). Box 3.2 outlines the relationship between these two scales.

It is more usual to obtain an absorption spectrum by means of a spectrophotometer, which is a more precise instrument than a colorimeter.

**Box 3.2 Absorbance and transmittance**

The laws governing the transmittance of light through a sample are inverse relations where doubling of one of the variables produces a halving of the other. A graph of this type of relationship is a hyperbola, as shown in diagram (a). This is because transmittance measures the light that **has not** interacted with the sample. In analysis we need a measure of how much light **has** interacted with the sample – the absorbance. When the laws are expressed in terms of the absorbance of the sample, a 'direct proportion' results. This is more useful because doubling of one of the variables produces a doubling in the other. A graph of this sort of relationship is a straight line through the origin and can give much greater precision than a curved graph.

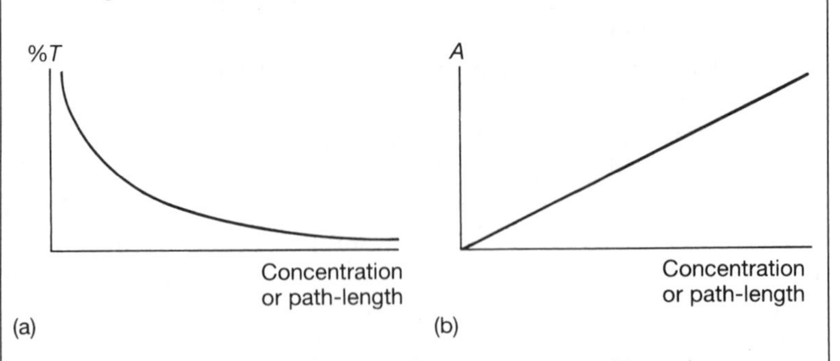


The relation between absorbance and percentage transmittance is a mathematical one. If we put $I_o$ for the intensity of light entering the sample and $I_t$ for the intensity emerging then the percentage transmittance is defined as:

$$\%T = (I_t/I_o) \times 100\%$$

The absorbance scale is derived from it thus:

$$A = \log(100/T\%)$$
$$= \log 100 - \log T\%$$
$$\text{so } \mathbf{A = 2 - \log T\%}$$

However, for practical purposes, we need not consider the relation further at this stage since the read-out on most modern colorimeters and spectrophotometers either gives the two scales (absorbance and transmittance) side by side, or the read-out can be switched from one to the other as required.

When absorbance is plotted against wavelength, the graph produced is known as the **absorption spectrum** of the sample. The lower portion of Figure 3.2 shows the absorption spectrum of chlorophyll; compare it with the absorption bands in the upper portion. For each absorption band there is a wavelength which gives maximum absorption in that region. This is known as the $\lambda_{max}$ (pronounced 'lambda max') of the absorption band and has a characteristic value for the sample. For chlorophyll the two values of $\lambda_{max}$ are 420 nm and 670 nm.

Figure 3.3 shows further typical absorption spectra. Try to correlate the shapes of the graphs with the colours of the substances.

Methylene blue is a dye often used in experiments to determine the mechanism of respiration. Can you work out what its absorption spectrum would look like, and an approximate value for its $\lambda_{max}$? The next few paragraphs should help.

### 3.1.3 Colorimetry

The technique of colorimetry is quite simple and is widely used in biological analysis. A typical colorimeter is shown in Figure 3.4. Although there is a wide variation in the look of instruments from different manufacturers, the basic components are the same.

Before you make any measurements, you must select an appropriate wavelength. Remember that it will be in the part of the spectrum that you do **not** see (i.e. the colour that is being absorbed by the sample). For example, if a sample is yellow (such as carotene) then it is transmitting in the yellow–red region of the spectrum, but it is absorbing in the blue region. Very basic colorimeters use coloured filters to select the wavelength and for a yellow sample you will need a filter that looks blue (one that transmits blue light). As a general rule, the colour of a filter must be complementary to that of the sample.

Many modern colorimeters allow you to set the wavelength more precisely than by using a colour filter and you should

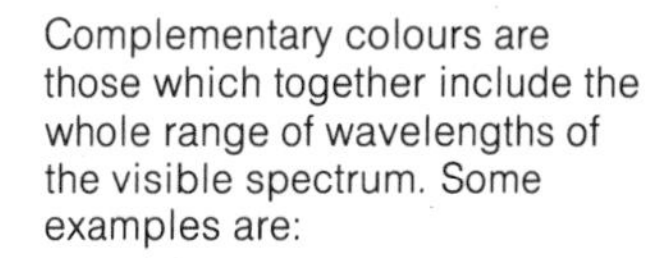
Complementary colours are those which together include the whole range of wavelengths of the visible spectrum. Some examples are:

- a yellow filter (transmitting green, yellow, orange and red) for a blue solution;
- a green filter for a magenta solution (purple and red wavelengths);
- a red filter for a cyan (blue-green) solution.

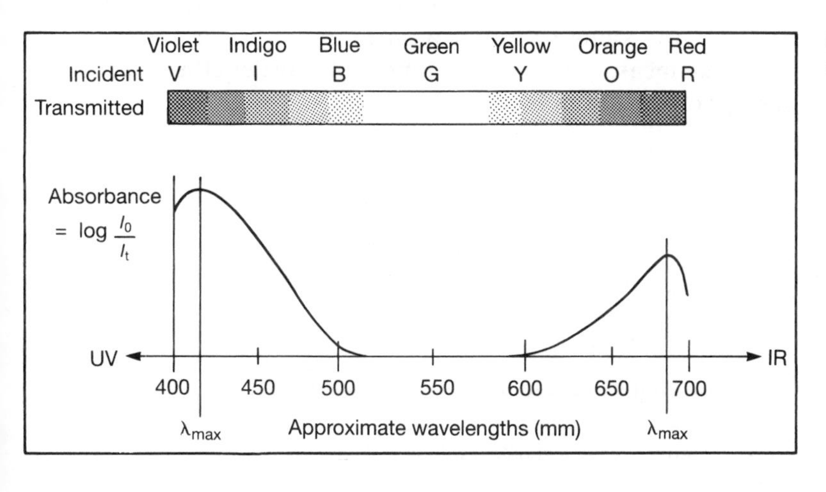


**Figure 3.2** The absorption spectrum of chlorophyll.

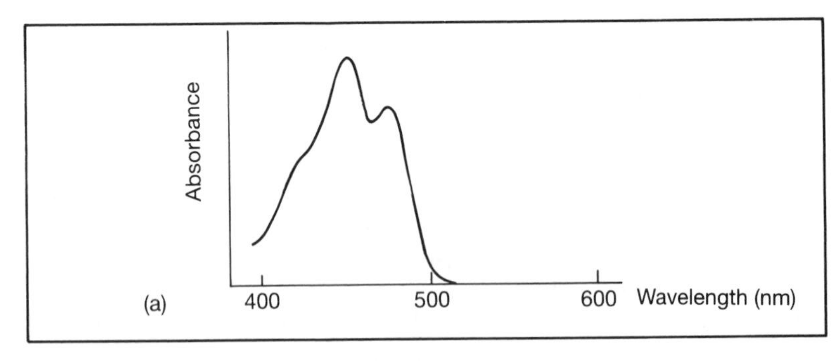


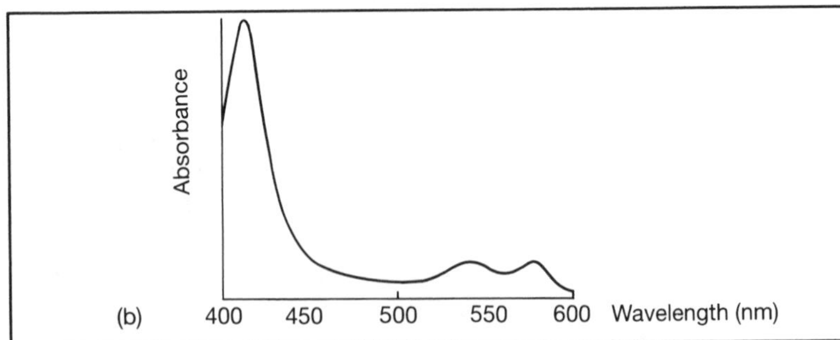


**Figure 3.3** Typical absorption spectra of coloured biological substances: (a) $\beta$-carotene; (b) oxyhaemoglobin.

choose the $\lambda_{max}$ value for the sample (see Box 3.3). But how do you know what that value is? With a little experience, you can make a reasonable guess. From what we said in the previous paragraph, you would expect $\lambda_{max}$ for carotene to be somewhere near 500 nm, the wavelength of blue light, so you could set the colorimeter to 500 nm. However, this is not very reliable (look again at the spectrum of carotene in Figure 3.3) and for most analytical work you will need a more accurate value of $\lambda_{max}$ than an informed guess. You can often find the value in a reference book or an instruction sheet, or you may have a copy of the absorption spectrum such as in Figure 3.3. However, if you do not have the value or if you wish to find $\lambda_{max}$ for yourself, you can make measurements at a number of wavelengths and plot the absorption spectrum (see Section 3.3).

**Box 3.3 Why choose $\lambda_{max}$ for colorimetry?**

Examination of an absorption spectrum such as in Figure 3.3 shows us ranges of wavelengths where there is no absorbance. If we were to use any of these wavelengths for that particular sample we would obtain a zero reading even if the sample were quite concentrated. We must use regions of the spectrum where there is measurable absorbance.

**But why should we use the particular wavelength $\lambda_{max}$?**

One reason is that $\lambda_{max}$ gives the highest reading in that range of wavelengths and consequently the instrument will be at its most sensitive. Another reason is that the calibration of the instrument is most reliable when $\lambda_{max}$ is used. See Section 3.1.3 on Beer's law and deviations from it.

**Do we have to use exactly $\lambda_{max}$?**

No, because the shapes of the absorption spectra show that there is always at least 10 nm around the $\lambda_{max}$ value where there is very little change in the absorbance reading. As long as the instrument is set to $\lambda_{max} \pm 5$ nm satisfactory results should be obtained.

(At this point you might be asking how we know that the wavelength scale of the instrument is accurate? If it is not, the actual wavelength being used might be more than 5 nm away from $\lambda_{max}$, which would give significant differences from the most accurate results. The wavelength accuracy of the instrument does have to be checked. See Section 3.3.1.)

When you use a colorimeter, such as in Figure 3.4 (p. 48), you have to:

- make a zero transmittance setting for the instrument (only at the beginning of the experiment);
- select the appropriate wavelength or insert the appropriate colour filter (only at the beginning or when you change to a different colour);
- insert a blank and set the absorbance reading to zero (again, only at the beginning or when you change to a different colour);
- replace the blank with a sample and record the reading shown on the absorbance display.

The usefulness of colorimetry in analysis is that the absorbance values are proportional to the concentration of the substance in the sample. Box 3.2 gave an indication of how absorbance is derived from instrumental considerations. We will now relate it to the experimental details. In all analytical work, knowledge of concentrations is important and colorimetry is a method of determining concentrations which is simpler and quicker than many methods.

The law that is the basis of such measurements is a combination of two observations.

1. **The greater the path-length (the distance the light travels through a solution), the greater the proportion of light that is absorbed.** More precisely, we find that the **absorbance ($A$)** of a solution is **directly proportional** to the **path-length ($b$)**.
2. **The greater the concentration of the sample the greater the proportion of light absorbed**. The **absorbance** of a solution is **directly proportional** to the **concentration ($c$)** of the absorbing substance in the solution.

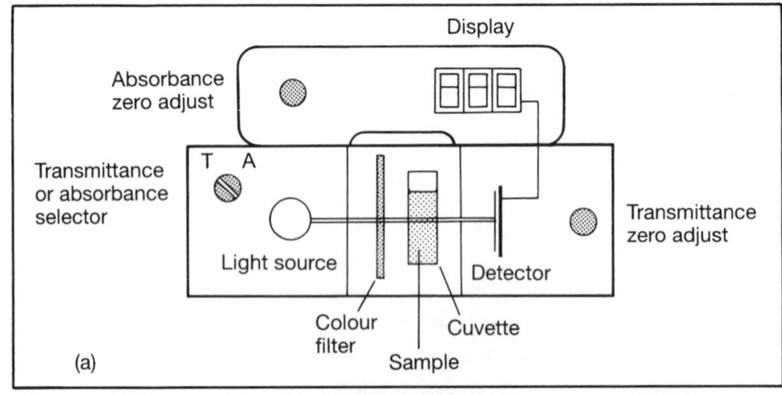

**Figure 3.4** A simple colorimeter: (a) the components; (b) a commercial instrument (courtesy Jenway, UK).

Beer, Lambert and Brouguer were early workers in the field of absorption of light and their names have become associated with the absorbance law. It is not clear, however, whether any of them actually enunciated the law or any part of it as such! A recent article (G. B. Levy, *International Laboratory*, June 1993, pp. 4–5) catalogues the few details that are known for sure and comes to the conclusion that the law gradually evolved into its present form. So, while we seem to honour Beer and Lambert more in the UK, other countries such as France and the USA sometimes acknowledge the contribution of Brouguer in their naming of the law.

These two laws can be combined into one simple relationship, known as the **Beer–Lambert law** or occasionally as Beer's law:

$$\textbf{absorbance} = \textbf{constant} \times \textbf{path-length} \times \textbf{concentration},$$

or, using symbols:

$$A = abc$$

where the constant, $a$, is the **absorptivity**, if the concentration has been measured in g $dm^{-3}$.

If, however, the concentration was expressed in mol $dm^{-3}$ the constant would be the **molar absorptivity** and the symbol would probably be written as $\varepsilon$. The Beer–Lambert equation would then be:

$$A = \varepsilon b C_m$$

The absorptivity of the sample has a characteristic value for that sample. It does not depend on the size of the sample but it does vary with the wavelength, so absorptivity values must have their wavelength specified. If you know the absorptivity of a substance you can easily find its concentration by measuring its absorbance and using the calculation $c = A/ab$.

For example, the absorptivity ($a$) of the copper(II)–protein complex which is formed in the biuret reaction is $0.05\,cm^2\,mg^{-1}$ at 545 nm. If you had a solution of a protein treated in this way whose absorbance ($A$) was 0.33, when measured in a 1 cm cell, you would calculate that:

$$c = \frac{A}{ab} = \frac{0.33}{0.05\,cm^2\,mg^{-1} \times 1\,cm} = 6.6\,mg\,cm^{-3}$$

Some absorptivity values can be found in biochemistry textbooks and tables of them are in reference books such as the *CRC Handbook of Biochemistry and Molecular Biology*, edited by G. Fasman (3rd edition, 1976, CRC Press). However, do remember that calculations using a single measurement are less reliable statistically and you should use the mean of several measurements.

The units for absorptivity are somewhat unusual and depend on the concentration units used. If the mass concentration is used (g $dm^{-3}$) and the path-length is in cm, then absorptivity has the units $dm^3\,g^{-1}\,cm^{-1}$. Similarly, with molar concentrations, the units are $dm^3\,mol^{-1}\,cm^{-1}$. Unfortunately, you may find them written as $l\,g^{-1}\,cm^{-1}$ and $l\,mol^{-1}\,cm^{-1}$ and even $M^{-1}\,cm^{-1}$ for this last. None of these conforms to SI conventions. When such units are converted to SI they become $cm^2\,mg^{-1}$ and $cm^2\,mmol^{-1}$ respectively. It is probably not worth trying to interpret what the units mean in a physical sense but you should make sure you recognize whether the concentration was in terms of mass or moles.

### 3.1.4 Using a calibration graph: the standard curve

An alternative method of finding concentrations does not require knowledge of the absorptivity value. You measure the absorbances of a range of known concentrations so that you can calibrate the instrument. This is usually done by drawing a calibration graph (often called a 'standard curve', see Section 1.6). The readings from the instrument are plotted against the values of the concentration. Then your test solutions are put into the colorimeter and their readings taken. These readings are converted into the corresponding values of concentration using the calibration graph.

For example, the following readings were obtained for standard solutions of egg albumin, treated with biuret reagent and measured at 545 nm in the same manner as for the previous example.

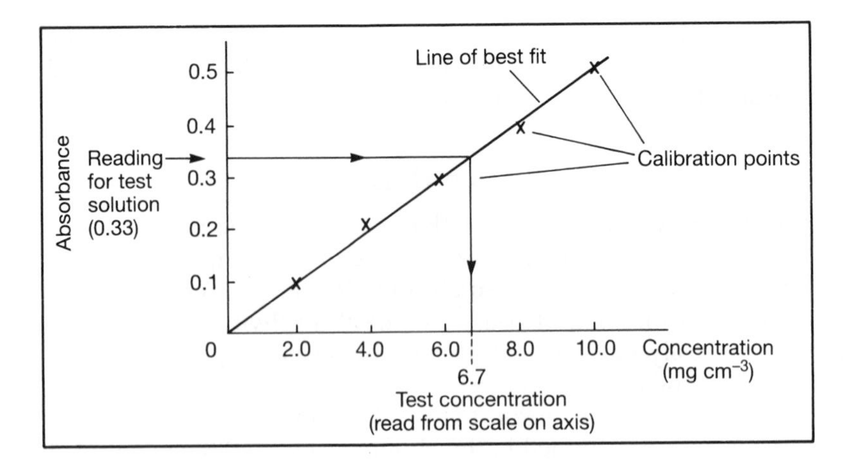


**Figure 3.5** Calibration graph for colorimetry.

| *Concentration (mg $cm^{-3}$)* | *Colorimeter reading* |
|---|---|
| 2.0 | 0.10 |
| 4.0 | 0.21 |
| 6.0 | 0.29 |
| 8.0 | 0.39 |
| 10.0 | 0.50 |

From these figures we can draw the graph shown in Figure 3.5. (Assume that zero absorbance has been correctly set using a proper blank.)

Then if we have a test sample whose concentration we wish to find which gives a reading of 0.33, we can determine it to be 6.7 mg $cm^{-3}$.

The slight difference between this result and the previous one could be ascribed to the difference in precision of the two methods.

## 3.2 The basis of spectroscopy

### 3.2.1 The electromagnetic spectrum

In the previous section we considered the visible spectrum, but we implied that there were at least two other regions of radiation: the UV (ultraviolet; shorter wavelengths than visible) and the IR (infrared; longer wavelengths than visible) (Figure 3.2). In fact, the complete range of wavelengths is much more extensive than UV, visible and IR. Even shorter wavelengths include X-rays, $\gamma$-rays (gamma-rays) and cosmic rays, and even longer wavelengths include microwaves and radiowaves. They are all forms of **electromagnetic radiation**, and in a vacuum they all travel with the same velocity – the speed of light. The various ranges are defined

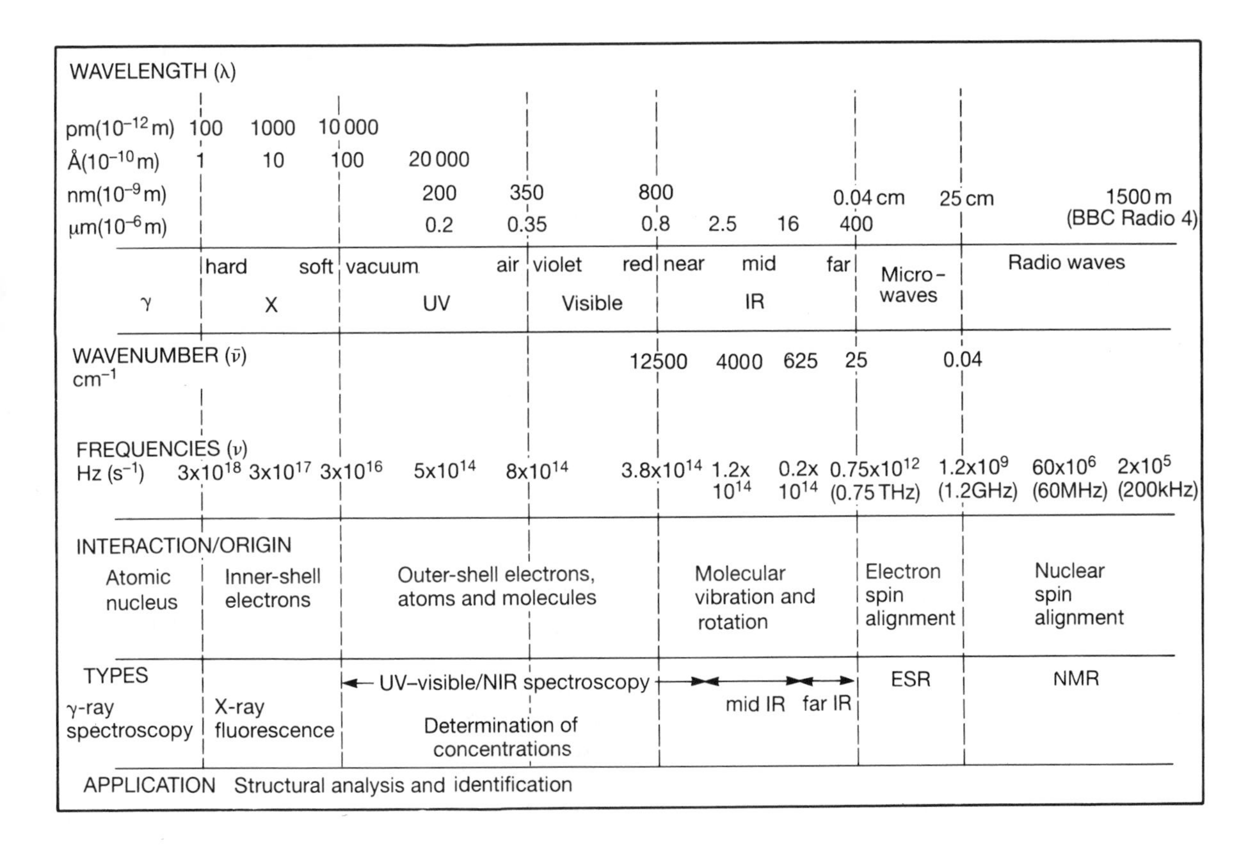


**Figure 3.6** The electromagnetic spectrum.

by the different methods of detection and of generation that they require, as well as having somewhat different interactions with matter. You will find it useful to recognize the ranges of wavelengths of the regions. These are given in Figure 3.6 along with other details to give an overall view of the extent of spectroscopy.

### 3.2.2 Types of spectra

Electromagnetic radiation is useful to the analyst because it interacts with atoms and molecules. In this context, the interaction is an exchange of energy and the two principal processes are **absorption** and **emission**. When radiation is absorbed, energy passes from the radiation to atoms and molecules which become excited. Of course, atoms and molecules can be excited by other means: the heat of a flame or the energy of an electric discharge. Once atoms and molecules have become excited they may lose that energy by various means, including conversion to heat or sound, but in spectroscopy we are principally concerned with the loss of energy by emission of radiation. So far we have only considered absorption spectra but we can mention some familiar everyday examples of emission spectra. The ordinary filament lights at home or in a car headlight produce white light together with a lot of IR (they get very hot) but little UV and no other radiations. The radiation from an electric fire is similar but there is much less visible and no UV radiation. Yellow street lamps use energized sodium atoms to produce light, with little UV and IR. Neon tubes work similarly. Fluorescent tubes produce UV radiation but this is converted to visible light by the fluorescent coating on the inside of the tube. Light-emitting diodes were used in early calculators. Biologically, light is emitted by the firefly, and many biological molecules can fluoresce – quinine and NADH, for example.

As with absorption spectroscopy, emission spectroscopy can be used both qualitatively and quantitatively.

### 3.2.3 Band spectra and line spectra

The absorption spectra shown in the section on colorimetry are known as **band spectra** since they show bands of absorption with indistinct edges and gradual changes of intensity. This type of spectrum is produced when the sample contains molecules. The emission spectra of filament lamps are also band spectra since a wide range of wavelengths is produced.

However, when the light from a sodium lamp or from a neon tube is analysed the spectra are found to consist of separate narrow lines (**line spectra**). This is because the light is being emitted by free atoms. Each line corresponds to an emission at a particular wavelength and each element has a characteristic pattern of lines for its atoms. Free atoms usually require gaseous systems and high temperatures, consequently they are rare in biological samples. It is also possible to have absorption spectra for free atoms if they are not excited. You can see absorption lines in the spectrum of sunlight (Plate 2).

These lines were first noticed by J. Fraunhofer (1787–1826), a German physicist who worked extensively in optics. The white light generated by the sun shines through the atomized gases of its rarified outer atmosphere and so some wavelengths have been absorbed before the sunlight reaches us.

### 3.2.4 Energy levels in atoms and molecules

We now need to consider why atoms and molecules absorb and emit some wavelengths and not others. The explanation will also show us why the wavelengths are characteristic (in other words, typical of a particular atom or molecule).

In the world of our normal experience we find that the amount of energy that a system or an object may possess can vary in a continuous manner. That is, its value can vary smoothly between upper and lower limits and there are no gaps in the range. The kinetic energy of a car varies continuously as its speed changes; the amount of heat in a beaker of water varies continuously as its temperature rises or falls. However, at the atomic level, energy is discontinuous! Atoms and molecules may possess certain levels of energy, but not amounts in between these levels. Atomic and molecular energies are thus said to be **quantized**. An anology from the normal world is a ladder. The potential energy of a person on the ladder is defined at definite levels by the rungs. The climber cannot stand at a level between the rungs.

The energy levels available to atoms and molecules correspond to:

- the various possible arrangements of the electrons in the orbitals of atoms or molecules, **electronic energy levels**;
- various amounts of vibration of covalent bonds in molecules (there is no corresponding vibration in free atoms), **vibrational energy levels**;
- various amounts of rotation of molecules (again, there is no corresponding motion in free atoms), **rotational energy levels**.

### 3.2.5 The relationship between energy and wavelength of radiation

We have just considered the quantization of the energy levels of atoms and molecues. We must now pursue a similar consideration for electromagnetic radiation. Each wavelength is associated with quanta, known as **photons**, of a particular energy (see Box 3.4). No two wavelengths have photons of the same energy. Also, because the relationship between the energy and the wavelength is inverse (or reciprocal):

- **shorter wavelengths** are associated with photons of **higher energy**;
- **longer wavelengths** are associated with photons of **lower energy**.

This is why UV (shorter wavelength/higher energy) can cause photochemical reactions; the photon energy is sufficient to break some chemical bonds and initiate reactions. However, with IR (longer wavelength/lower energy) the radiation does not induce abnormal chemical changes; these photons are not energetic enough to break chemical bonds.

**Box 3.4 Quantization and the duality of nature**

It became clear at about the turn of the last century that some of the interactions between light and matter could not be adequately explained by the wave theory of light. In fact, some predictions of the theory were quite wrong. The energy distribution in the spectrum of radiation from a black object, and the so-called photoelectric effect, led to the proposition that electromagnetic radiation must in some way be particulate, or quantized. This may seem surprising since the wave nature of the radiation gives no indication that it also has properties which are discontinuous.

The quanta of radiation are often described as 'particles of light' or 'packets of energy' but no such description is entirely adequate. They have been given the name photons. They have no mass but they do have a definite energy.

When a beam of monochromatic light is characterized by its wave properties, such as by interference effects, its wavelength can be measured. When the **same** beam is characterized by its quantum effects the energy of its photons can be measured. There turns out to be a simple relationship between this energy and the wavelength:

energy (J) of photon = a constant (J s) (known as Planck's constant; see below) × frequency (Hz or $s^{-1}$) of the radiation    In symbols: $E = h\nu$

but since    $\text{frequency } (s^{-1}) = \dfrac{\text{velocity } (m\,s^{-1})}{\text{wavelength } (m)}$    In symbols: $\nu = \dfrac{c}{\lambda}$

so    $\text{energy } (J) = \text{constant } (J\,s) \times \dfrac{\text{velocity } (m\,s^{-1})}{\text{wavelength } (m)}$    In symbols: $E = h \times \dfrac{c}{\lambda}$

Once it became accepted that radiation had this dual nature it was soon asked whether particles with mass, such as the electron, could show wave-like properties. The diffraction of a beam of electrons showed that they could, and in an electron microscope the wave properties of the beam of electrons allow very high magnifications to be obtained. Nature has a dual character.

Since then spectroscopy has advanced enormously and has given rise to a much greater understanding of the natural world, but at the cost of the difficulty of holding two very different concepts, particles and waves, in mind at the same time!

Max Planck (1858–1947), a German physicist, was prominent in the fields of electromagnetic radiation and quantum mechanics. He is noteworthy for establishing this link between the two rather disparate theories.

### 3.2.6 The absorption and emission processes

- **Absorption.** When an atom or molecule becomes excited there is a transition from a lower to a higher energy level. Thus there is a precisely defined amount of energy required for the process and the energy of the photon being absorbed must match this exactly. Thus the wavelength being absorbed is defined.
- **Emission.** When excited atoms or molecules lose their energy, the transition is from a higher to a lower energy level and, again, the difference between the energy levels produces a particular wavelength.

The solvent affects the spectrum of molecules in solution, producing a smoothing of the shape of the absorption band. Solvents which interact strongly with the molecules produce more smoothing than those that do not, so you should always specify the solvent used when reporting these results.

We have noted previously that free atoms have electronic energy levels only. These are widely spaced with different gaps between them and so the absorption and emission spectra of atoms are distinct lines. Molecules, however, have electronic, vibrational and rotational energy levels with much closer spacing of the vibrational and rotational levels. There are many possibilities for transitions with a closely similar energy difference so the spectra could be thought of as being many lines close together. For a given atom or molecule the energy levels are effectively fixed, and so its spectral properties are characteristic.

### 3.2.7 Fluorescence spectra

You may have come across some materials that glow in some way when UV light is shone on them. These fluorescent materials can be used to good effect in stage shows, and they are used in clothing materials where visibility is important or in detergents to make white materials look 'even brighter' after washing. All of these materials absorb at shorter wavelengths and emit at longer wavelengths. In ordinary absorption spectroscopy a photon is absorbed and the energy acquired by the molecule is either transferred to other molecules in the form of heat, or it is re-emitted as a photon so rapidly (less than $10^{-15}$ s) that there is no time for other energy levels to be involved. Hence the wavelength of the radiation is unchanged. This re-emission is known as 'resonance emission'. In fluorescence the photon absorbed excites the molecule both electronically and vibrationally. At this stage the electronic excitation is retained long enough (approx. $10^{-6}$ s) for collisions to take place and the vibrational energy to be passed on to other molecules without emission of radiation. When the molecules lose their electronic excitation the emitted photons have less energy than those absorbed, and the emitted radiation has a longer wavelength.

## 3.3 Ultraviolet–visible absorption spectrophotometry

The traditional design of a spectrophotometer uses a monochromator to disperse light from the source into its constituent wavelengths and selects a narrow range to pass through the sample. We will start by considering single-beam spectrophotometers, which are closest to colorimeters. Much of what we describe here will be relevant to more elaborate instruments described later. The designs vary according to how the manufacturer thinks an instrument of given specification can be economically produced. In particular, the geometry of the light path changes a lot from one instrument to another, but this does not hinder your use of them. The basic principles are the same.

### 3.3.1 Single-beam instruments

The main components of a single-beam spectrophotometer are indicated in Figure 3.7 and the following particular comments should be read with reference to that figure.

*Colorimeters.* The components of a colorimeter are cheaper and of more limited performance than those in spectrophotometers,

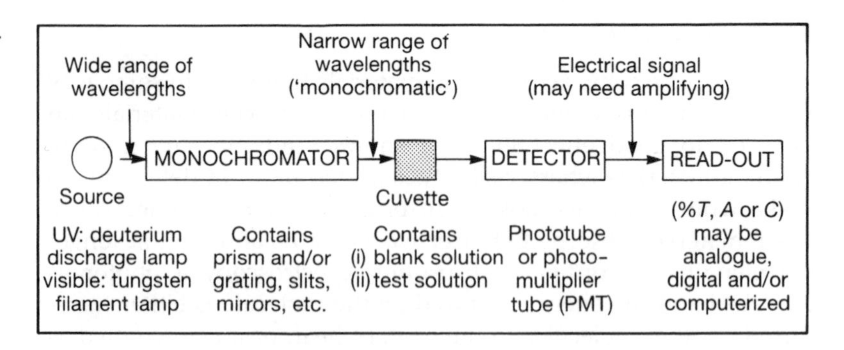


**Figure 3.7** Components of a single-beam spectrophotometer.

but give adequate precision for many routine applications. These instruments can be battery powered and made small enough to be portable, so they can be used in field-work for environmental monitoring, water and soil testing. More advanced instruments may use interference filters, which give better wavelength selection than colour filters.

*Spectrophotometers.* These are laboratory instruments which can give high performance throughout the UV and visible range (some models are restricted to the visible range to give a cheaper alternative). They use a grating monochromator and may have adjustable slits which allow the spectral bandwidth (the degree of monochromaticity) to be matched to the needs of the sample (see Box 3.5). The sample compartment has room for several types of sample holder, single cuvette, multiple cuvette, thermostatically controlled and flow-cell, among others.

**Box 3.5 Bandwidths**

Any wavelength selection system, even a laser, produces a range of wavelengths. We may talk of monochromatic radiation when the range is very narrow, but truly monochromatic radiation is not possible (it would violate Heisenberg's uncertainty principle!). Thus, in addition to the nominal wavelength, we should refer to the bandwidth of the radiation which states the numerical difference between the shortest and the longest wavelengths that are present to a significant extent in the beam. You may come across the term with respect to frequencies and wavenumbers as well as wavelengths.

The importance of the bandwidth is that a machine will not give an accurate representation of an absorbance band unless the spectral bandwidth is less than about one-tenth of the width of the absorbance band. Without this condition, narrow, closely spaced absorption bands will not be resolved (i.e. shown as separate bands).

There is little problem when using UV–visible spectrophotometers for biological molecules since their absorption bands are usually quite wide, but it does become more noticeable in colorimeters, where the filter system produces a fairly wide spectral bandwidth. In mid-IR work narrow absorbance bands may also be a problem.

| *Wavelength selector* | *Spectral bandwidth (nm)* | *Typical absorbance bandwidth (nm)* |
|---|---|---|
| Colour filter | approx. 50 | |
| Interference filter | 5–10 | 100 |
| Monochromator | 0.5–4 | |

*Operating procedure for a single-beam spectrophotometer*

Figure 3.8 shows a typical layout of a single-beam spectrophotometer. The reference numbers correspond to the steps given below. You will have to check that you can find the corresponding controls on your instrument. The most recent machines have a keypad and screen with on-screen instructions – follow them, of course, but you will find that the steps are much the same as given below.

**Preliminaries**

1. Switch on and allow the instrument and lamps to warm up. This will take about five minutes and, during this time, modern machines will go through an automatic setting-up procedure.
2. Set zero transmission (NOT absorbance) if not already done automatically. This is to allow for any signal that the detector produces even if it is not receiving any light.
3. Select required wavelength. You may have to do this via a keypad.

**Calibration**

4. Select absorbance. Some machines will offer a choice of ranges (use 0–1 if you are in doubt). Also, some machines offer

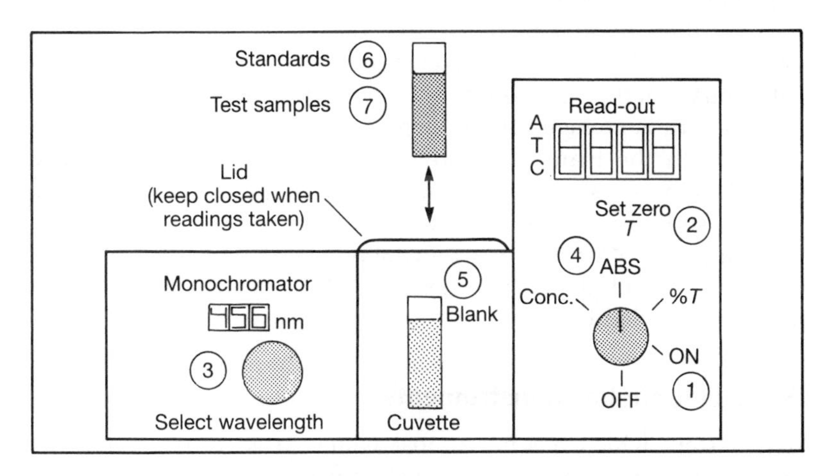


**Figure 3.8** Operation of a single-beam spectrophotometer.

calibration in terms of concentration, but you need to be sure that Beer's law holds for the range of concentrations you are likely to encounter so you will have to do an absorbance calibration first.
5. Put the blank solution in the sample holder and adjust to give zero absorbance reading. This adjusts the amplifier so that any absorbance in the blank and any scattering from the cuvette is allowed for. Check in Chapter 2 if you are unsure what is an appropriate blank.
6. Put the standards in and record absorbances. Do the lowest concentration first, then in order of increasing concentration. Use same cuvette for each and rinse with next solution to be used before measuring. Check the linearity of the calibration; if there is any distinct curvature use a more dilute set of standards.

**Test samples**

7. Insert test samples and take readings without further adjustment. Use same cuvette as for standards. If any readings fall outside the calibrated range, they must be diluted by a known factor and re-read.
8. If a change of wavelength is required, go back to step 3. This will be required if you are scanning through a range of wavelengths, which makes it a rather laborious procedure.

*Deficiencies of the single-beam system*

There is an inherent weakness in the single-beam system, which is that if the source intensity varies between the time of setting the blank and the measurement of the sample, then the reading will be inaccurate. Some manufacturers incorporate an extra detector to monitor and adjust the source intensity to increase the reliability of the readings.

A further disadvantage of the design is that the process of scanning through the wavelengths (perhaps to find $\lambda_{max}$) is laborious whether the source is stabilized or not. Every time the wavelength is changed the system has to be reset to zero absorbance with the blank. The advent of cheap computer memory circuits allows a scan of the blank or background to be done and stored, so that it can be subtracted from the sample data while it is being run. Since there is inevitably a time delay between the blank scan and that of the sample, these instruments need a highly stable light source.

### 3.3.2 Double-beam instruments

The system which has been most widely used to overcome the deficiencies of the single-beam system uses a double beam, in the

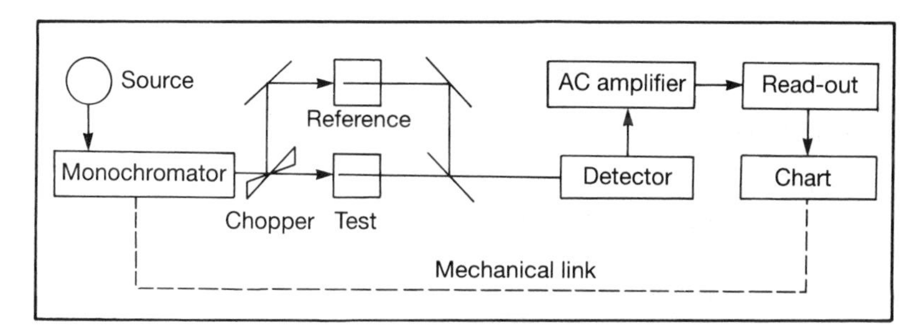


**Figure 3.9** Block diagram of a double-beam spectrophotometer.

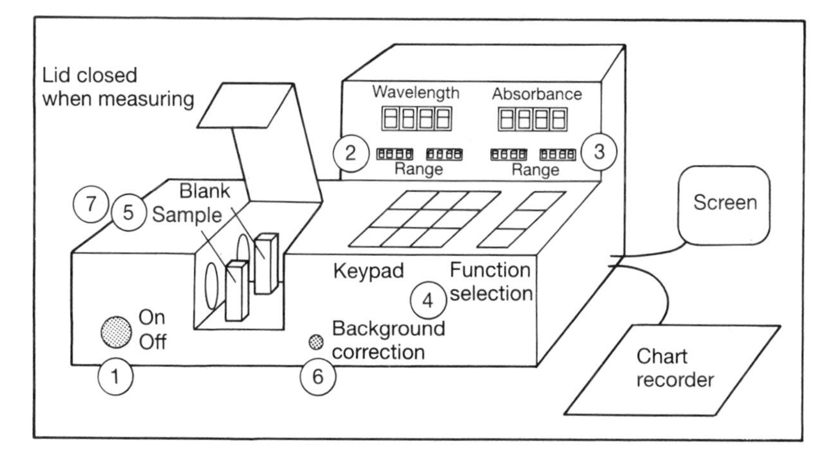


**Figure 3.10** Operation of a double-beam spectrophotometer.

sense that the beam is split so that part goes through a cell containing the sample and part through another cell containing a blank, as shown in Figure 3.9. The beam splitting is done by a rotating segmented mirror (a 'chopper') which alternately directs the beam through one cell and then the other. The two beams are recombined and produce an alternating signal at the detector which can be more accurately amplified and processed to give the **difference** between the two beams. Thus there is continuous compensation for any absorbance in the blank. It is obvious that the double-beam system needs more components than the single-beam and it is more expensive and can be more susceptible to breakdown. Nevertheless, it has done good service for many years. Whether it will be superseded by the computerized single-beam system mentioned above remains to be seen!

*Operating procedure for a double-beam instrument*

Figure 3.10 shows a stylized layout of a double-beam spectrophotometer. Check that you can find the corresponding controls on your instrument. The most recent machines have a keypad and screen with on-screen instructions; follow them, of course, but you will find that the steps are much the same as given below.

The reference numbers in the figure correspond to the following steps.

**Preliminaries**

1. Switch on and allow time for warm-up, and for initiation protocol on microprocessor machines.
2. Select the wavelength range to be used (or a single wavelength).
3. Select the absorbance range to be used (e.g. 0–1 preliminary work or 0–0.1 for very dilute solutions. You may need to do some dilution of more concentrated solutions since the instrument is unlikely to give reliable readings of absorbances greater than 3.
4. Leave other parameters at default settings unless you have specific information about the need for other settings.
5. Put blank solutions in **both** sample and reference beams. Note that the sample beam is always toward the front of the machine and the reference toward the rear.
6. Perform blank scan/background correction (or autozero for single wavelength).

**Scanning**

7. Put a standard or a test sample in the sample beam; leave the blank in the reference beam. Initiate the scan protocol. The absorption spectrum will appear on a chart recorder or on screen, or both; make sure you obtain a copy on paper for your records. You can determine $\lambda_{max}$ from the spectrum print-out, and some machines will do it from their control panel.

**Beer's law measurements**

8. Once you know $\lambda_{max}$, use that wavelength, and work through standards and test samples as for a single-beam machine.

### 3.3.3 Sample handling

*Cuvettes*

Almost all UV–visible spectroscopy is done with samples in solution. The containers for the solutions that are inserted into the spectrophotometer are called cuvettes or, sometimes, cells. A typical cuvette is about 4 cm deep and 1 cm across (this is the path-length). You use about 3 $cm^3$ of the solution for each measurement; it is better not to fill it completely. Most kinds have two opposite faces frosted, for handling. The other two are clear for transmission and should not be touched. You should rinse the cuvette with the next solution it is to contain, fill for measurement and check for bubbles. Then wipe the outside gently with lens

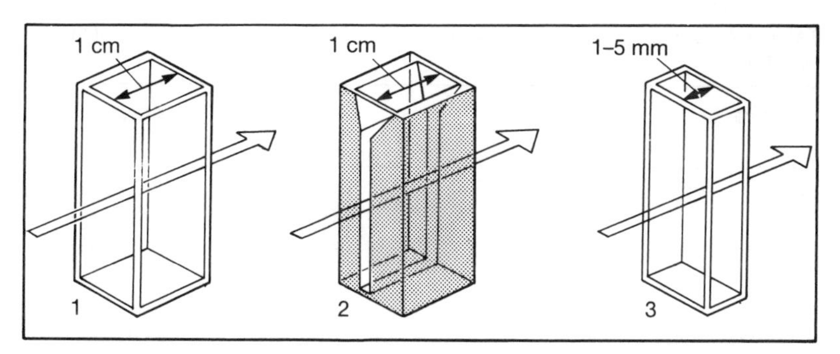


**Figure 3.11** Cuvettes for UV–visible spectrophotometry (courtesy Shimadzu Corp.). (1) A routine 1 cm cuvette; requires about 3 $cm^3$ of sample. (2) A low-volume 1 cm path-length cuvette; requires about 0.5 $cm^3$ of sample. (3) A short-path-length (1–5 mm) cuvette.

tissue before inserting it into the machine. Figure 3.11 shows a routine cuvette and some others for special applications.

The material of the cuvette must be appropriate. Plastic cuvettes are acceptable where high precision is not required, otherwise glass is used. Both can only be used for visible light as they absorb in the UV range. UV work requires quartz cuvettes, which are transparent to both UV and visible light. These must be handled with care as they are always made suitable for high precision.

*Solvents*

These must not interfere with the absorbance of the sample and so you should use a solvent only if its absorbance is less than about 0.3 in the region of interest. Most solvents are simple structures and they tend to absorb towards the short wavelength end of the UV spectrum. This produces a solvent **cut-off**, which is defined as the wavelength below which the solvent absorption (in a 1 cm cell) is greater than 0.3. Table 3.1 lists some values.

**Table 3.1** Solvent UV cut-offs

| *Solvent* | *Cut-off (nm)* |
|---|---|
| Acetone (propanone) | 230 |
| Acetonitrile (propanonitrile) | 190 |
| Carbon tetrachloride (tetrachloromethane) | 265 |
| Chloroform (trichloromethane) | 245 |
| Cyclohexane | 210 |
| Ethanol | 210 |
| Methanol | 210 |
| Toluene (methylbenzene) | 286 |
| Water | 191 |

**Table 3.2** Correlation of structure with $\lambda_{max}$ in the UV and visible regions of the spectrum

| *Structural group* | | $\lambda_{max}$ *(nm)* | *Absorptivity* *($dm^3 mol^{-1} cm^{-1}$)* |
|---|---|---|---|
| Ether | $-O-$ | 185 | 1 000[a] |
| Amine | $-NH_2$ | 194 | 2 800[a] |
| Nitrile | $-C{\equiv}N$ | 160 | Large[a] |
| Ketone | $>C{=}O$ | 195 | 1 000[a] |
| Ester | $-COOR$ | 205 | 50 |
| Carboxylic acid | $-COOH$ | 200–210 | 50–70 |
| Aldehyde | $-CHO$ | 210 | >5 000 |
| Polyene | $-(C{=}C)_2-$ | 220 | 21 000 |
| | $-(C{=}C)_3-$ | 260 | 35 000 |
| | $-(C{=}C)_4-$ | 300 | 52 000 |
| | $-(C{=}C)_5-$ | 330[b] | 118 000 |
| Benzene ring (band 1) | | 184 | 46 700[a] |
| Benzene ring (band 2) | | 202 | 6 900 |
| Benzene ring (band 3) | | 255 | 170 |
| Azo | $-N{=}N-$ | 285–400[b] | 3–25 |

[a] Much UV work is done at wavelengths greater than 200 nm so these absorption bands give rise to only a small amount of absorption in the normal working range. When UV absorption is used as the detection method in HPLC and the required wavelength for detection is less than 200 nm several commonly used solvents absorb significantly and may interfere with the detection.
[b] These bands stretch into the visible region and so give rise to coloured compounds even though $\lambda_{max}$ may be in the UV.

### 3.3.4 Typical uses of ultraviolet–visible spectrophotometry

1. Finding $\lambda_{max}$ perhaps to help distinguish between isomers.
2. Determination of the concentration of coloured substances at visible wavelengths such as haemoglobin and plant pigments.
3. Determination of colourless substances by reaction with a colour reagent: protein using the biuret reagent or sugars with DNSA (dinitrosalicylic acid).
4. Determination of colourless substances at UV wavelengths such as protein at 220 nm or acetylsalicylic acid (the active ingredient in aspirin tablets) at 296 nm.
5. Determination of two substances: phenacetin and caffeine as in a pharmaceutical preparation which used to be available or DNA and protein levels in extracts from particular cells.
6. Time curves for enzyme kinetics: if the absorbance of a substrate is changed by the action of an enzyme the rate of change can be determined by measuring the absorbance over a period of time.
7. The activation energy ($E_{act}$) is an important factor in chemical reactions and can be used as a measure of the effectiveness

of an enzyme such as invertase. It can be determined from measurements of the absorbance of substrates or products at different temperatures.

### 3.3.5 Recent advances in spectrophotometer design

*Diode-array detection*

In this system the light from the source is dispersed as before but then the whole spectrum of wavelengths falls on an array of detectors (light-sensitive diodes) so that effectively each wavelength is monitored simultaneously. The array is read into a computer memory and subsequently processed in any appropriate manner. The advantages are that scanning is extremely rapid and the optics are less complex. The system is particularly useful in high-performance liquid chromatography detectors (see Chapter 5).

*Open sample compartment system*

If the monochromator is put after the sample compartment (as it is in IR instruments; see Section 3.5) then it is no longer necessary to cut out extraneous light, and the sample is much more accessible. Rapid scanning can be done but since the sample is being irradiated with all the wavelengths including short wavelengths (high energy) there is a danger of decomposing your sample.

*Fourier transform (FT) system*

This is an alternative to scanning. It produces an interference pattern between the sample beam and a reference beam (the optics are very different from those of the normal double-beam system described above). The pattern is generated very quickly, but it has to be transformed mathematically into the normal form of the data. This requires an immense amount of computation but modern computers do it in about 30 s, so it is still a very rapid system. It does not, however, show a marked advantage over traditional systems for UV and is not widely used (in contrast to IR).

J.B. Fourier (1768–1830) was a French mathematician who devised a mathematical procedure for converting between functions of frequencies (corresponding to our normal spectra) and functions showing the variation in time or distance (corresponding to the interferometer patterns).

## 3.4 Near-infrared spectroscopy

Near-infrared spectroscopy (NIR) is a comparatively new technique in terms of routine analysis, although it has been used in some specialized studies over the past 40 years. It has now emerged as a widely applicable technique because of the availability of powerful microprocessors which can handle the rather cumbersome calculations needed to make it quantitative. You are likely to come across it in areas such as food quality control and monitoring of fibres.

NIR is the portion of the electromagnetic spectrum that is nearest to the red end of the visible region. It runs from about 800 nm to about 2500 nm (2.5 mm), where the mid-IR starts.

Look back to Section 3.2 and Figure 3.6 if you need to revise this topic.

The origins of the absorption bands in the near-IR are vibrations of the covalent bonds occurring at twice or three times the fundamental frequency, but combinations of bands of similar frequency produce further bands. Unfortunately, this results in no clear correlation between the spectrum and the structure of the molecules, unlike UV–visible and mid-IR spectroscopy.

### 3.4.1 The appearance and uses of near-infrared spectra

NIR spectra may be recorded in a similar manner to those of UV–visible spectra. A graph is produced of absorbance (or reflectance if the sample is opaque) against wavelength. More absorption bands are seen than in UV–visible spectra, but fewer than in mid-IR. For complex samples such as foodstuffs there may be few clear features (see Figure 3.12). When the nature of the samples is restricted to particular areas such as foodstuffs, fibres and so on, some bands can be ascribed to particular components of a sample: water, protein, oil, carbohydrate and the like.

The bands do lend themselves to quantitative determinations, using Beer's law principles. Calibration has to be done with particular care since other components in the sample may give some absorption at the wavelength of interest. This makes the position of the baseline rather variable. It is usual to use several wavelengths for a calibration, and 100 or so standards whose composi-

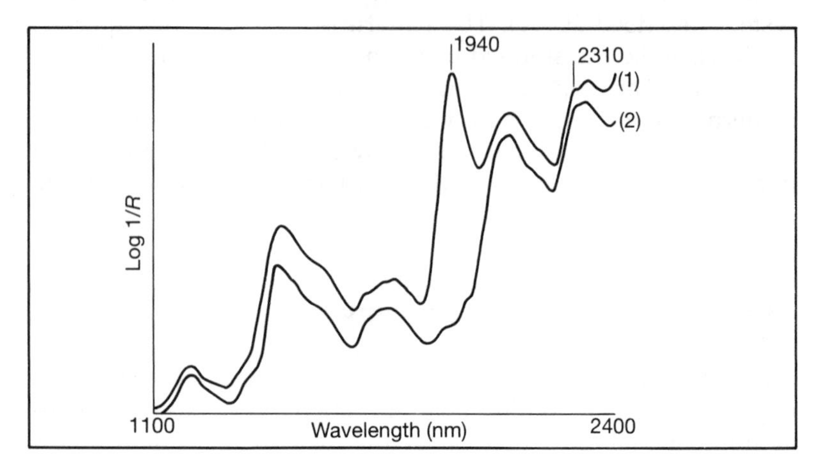


**Figure 3.12** Typical NIR spectra. The spectra are of a flour sample before (1) and after (2) drying. The absorption band at 1940 nm is due to the moisture; you can see that it is not present in spectrum 2. The wavelength 2310 nm is used as a reference to allow for the variable baseline in such spectra. The *y*-axis scale, in which *R* is the reflectance, is equivalent to the absorption scale used in transmission spectra. (Reproduced from B.G. Osborne, T. Fearn and P.H. Hindle, *Practical NIR Spectroscopy*, 2nd edn, published by Longman, 1993.)

tion has been determined by some other method. The data from the standards may be stored in the microprocessor in the instrument and used in programmed procedures for particular components. Calibration of near-IR machines is thus quite laborious and is often done by the manufacturer. Once a program has been set up, though, the technique is very easy to use; a test sample is measured at the preset wavelengths, the microprocessor solves a series of simultaneous equations and the result is displayed within about 30 s.

### 3.4.2 Near-infrared spectrophotometers

There are two main types of instrument: the pre-calibrated type, and the scanning type. The former is easier to use, and can be adapted for use with probes so that measurements of samples can be done on-line such as in a feed-pipe in a food-processing factory, or in places with awkward access such as sacks of flour. The latter is much more expensive, and is usually linked to a computer so that both qualititative and quantitative analysis can be done.

Some of the components of near-IR spectrophotometers are the same as are used in UV–visible instruments. Thus, the **source** of the radiation is a tungsten filament lamp (compare Figure 3.7).

Although we think of these lamps as producing visible light, they produce more energy in the NIR. You know how hot these bulbs get!

The **detector** is a lead sulphide device similar to that used in visible spectrophotometers for wavelengths greater than about 600 nm. The **optical path** in a scanning instrument is similar to those in UV–visible machines, but may differ considerably in

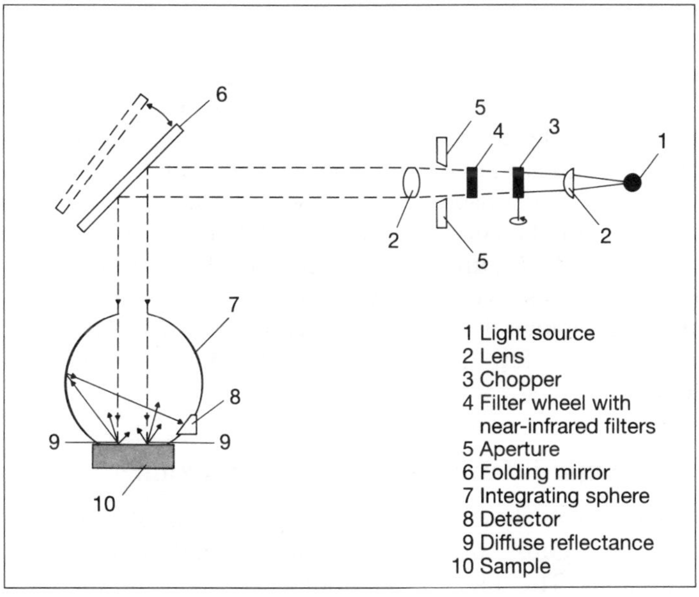

1 Light source
2 Lens
3 Chopper
4 Filter wheel with near-infrared filters
5 Aperture
6 Folding mirror
7 Integrating sphere
8 Detector
9 Diffuse reflectance
10 Sample

**Figure 3.13** Optical path for a pre-calibrated NIR instrument. (Courtesy Bran & Luebbe.)

the pre-calibrated type. Because samples are often opaque or insoluble, reflectance may be used instead of absorbance. For this the sample is placed in an **integrating sphere** which directs the reflected light onto a circular detector within the sphere, as shown in Figure 3.13. For calibration purposes the incident beam is directed onto the detector to obtain a reference value. Wavelength selection in pre-calibrated machines is by means of specially manufactured **interference filters**. In scanning machines a normal grating monochromator is used.

Samples may be in the form of powders, grains, liquids and emulsions. In a pre-calibrated instrument they are just placed in a tray and inserted into the machine. The actual procedure depends on the make of machine and so we cannot generalize here.

### 3.4.3 Applications (just a few of many)

Preset programs enable you to determine the amount of protein in a wheat sample, the percentage of glycol contaminant in wine, the level of fat in chocolate, the amount of unsaturation in a fat or the water content in coal or grain.

## 3.5 Mid-infrared spectroscopy

We now consider a part of the spectrum where the instrumentation has many similarities with UV–visible, but whose application and interpretation are rather different. Mid-IR spectroscopy has been used routinely as an analytical tool for about 30 years and has developed its own traditions. Until the recent advent of NIR spectroscopy, it was the only routine IR technique and so was (and still is) widely known as just IR spectroscopy. Its power lies in its ability to provide spectra which can be interpreted in terms of the molecular structure of the sample; it is less successful in providing quantitative data although some methods have been developed. It is less useful for biological samples than other spectroscopic methods since using it for solutions is not particularly straightforward. Routine use is with neat, dry, purified samples whose structures require investigating or confirming. Thus, samples of biological origin generally have to be extracted and processed; very few applications can be done on the original material.

### 3.5.1 The mid-infrared region of the spectrum

This region stretches from about 2.5 $\mu$m to about 16 $\mu$m wavelength (see Figure 3.6) but the usual designation is in terms of **wavenumber**. This unit can be understood as being the number of waves per centimetre and has the unit **cm$^{-1}$**. So we also refer to this region as being from 4000 cm$^{-1}$ to about 600 cm$^{-1}$. The wavenumber scale may seem a bit strange, but it is directly proportional to energy and this can help in interpreting mid-IR spectra.

You can use the formula wavenumber in cm$^{-1}$ = 10 000/wavelength in $\mu$m to convert wavelengths to wavenumbers. Check it with the figures for the ranges.

Mid-IR spectra usually contain 10–20 absorption bands and are thus much more complex than UV–visible and near-IR spectra. They often show distinctive patterns which aid the determination of molecular structures. The bands are characterized by their position in the spectrum (equivalent to $\lambda_{max}$ values), but are recorded in terms of the wavenumber.

Another idiosyncratic feature of mid-IR spectra is that they are usually presented as transmittance spectra, that is, the *y*-axis records percentage transmission so the baseline which corresponds to high transmittance (low absorbance) is at the top of the chart and the absorption bands point downwards and the tips of the peaks are near the bottom of the chart! A typical example is given in Figure 3.14.

### 3.5.2 Interpreting mid-infrared spectra

The absorption bands arise from the fundamental vibrations of covalent molecules which are associated with particular groups of

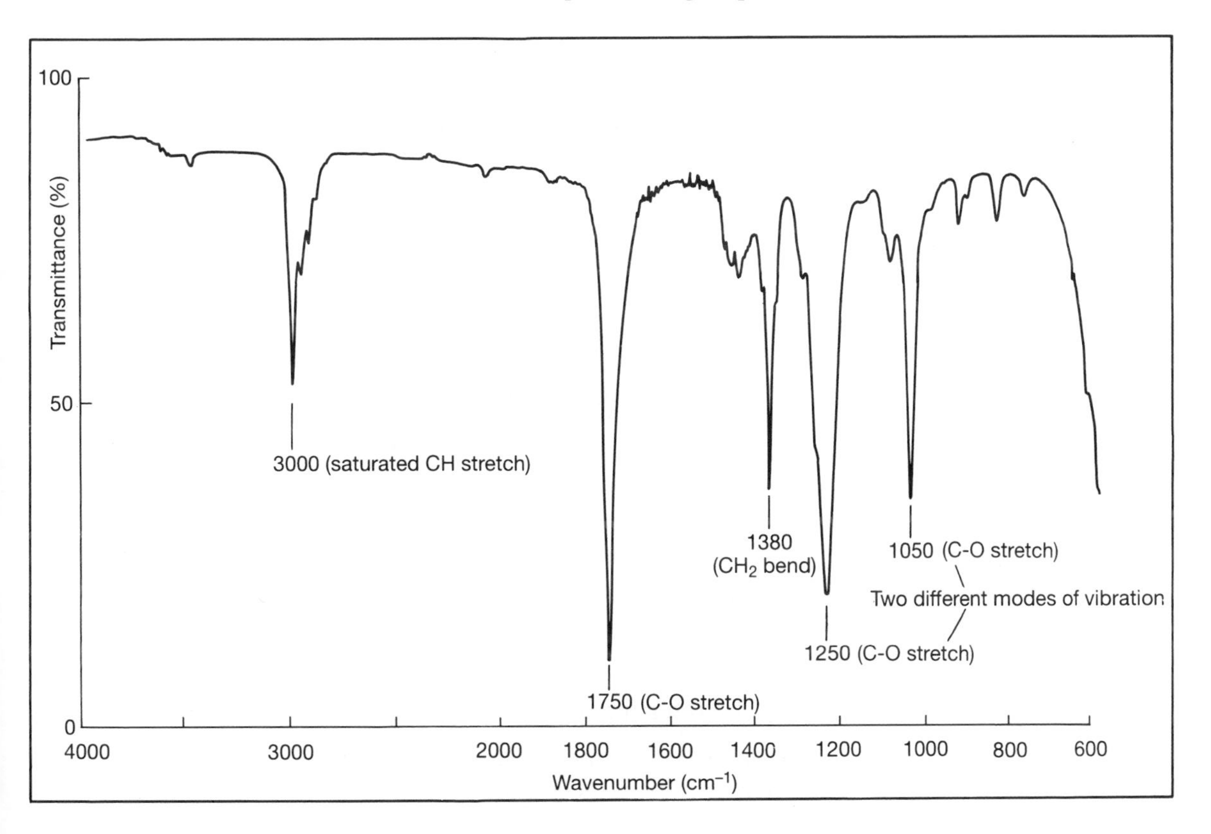


**Figure 3.14** IR spectrum of ethyl ethanoate (ethyl acetate).

**Figure 3.15** Simple correlation chart for IR absorptions.

Wavenumber ($cm^{-1}$)
4000 3000 2000 1700 1500 1000 600

Aliphatic CH |--| (str) |-CH-| (bend) |-----CH---------

Aromatic CH |--| (str) |--ring--| |--ring--|

|-C=O-| (str)

|--OH and NH--| (str) |---C-O---| (str)

IR absorptions are bands rather than lines because of the broadening effect of the rotation of the molecules.

atoms (functional groups), or of larger portions of the structure. There is a good analogy between the vibrations that a set of balls connected by springs could give rise to, and those of a molecule where the atoms are connected by covalent bonds. Thus there are vibrations of the atoms which cause stretching and bending of the bonds. Any functional group may be undergoing a combination of several of these motions, each with a different energy. When radiation of a particular wavelength (and thus of a particular wavenumber and energy) passes through a sample, it is only the vibration whose change in energy exactly matches that of the radiation that will absorb the radiation and become excited. Thus we can associate particular absorptions with particular groups, and we can make correlation charts (see Figure 3.15) of the position (wavenumber) of a band and the structural group from which it arises.

**Box 3.6 Rationalizing infrared correlation charts**

We can use the balls-and-springs analogy for molecular vibrations to help explain why different groups give rise to absorptions at different wavenumbers. First, we can use common experience to note that vibrations along the line of the spring (stretching and compression) take place more rapidly (that is, at higher frequency) than vibrations which bend the spring. Second, the masses of the balls affect the frequency of the vibration: more mass gives slower vibration (lower frequency) and less mass gives higher frequency. Finally, the strength of the spring is important: a stronger spring will give a higher frequency and so on. In molecular terms we have:

| **higher frequency and higher wavenumber** | **lower frequency and lower wavenumber** |
|---|---|
| produced by: | produced by: |
| stretching of bonds | bending of bonds |
| smaller atoms | larger atoms |
| stronger bonds | weaker bonds |

Try to rationalize the simple version of a correlation chart given in Figure 3.15 from these rules. If you are uncertain about the masses of any of the atoms you can find them in the periodic table (Appendix 2).

### 3.5.3 Determining molecular groups in samples

Notice that the ranges in which particular absorption bands appear are narrower in the left-hand half (4000–1400 $cm^{-1}$ approx.) of the correlation chart (Figure 3.15). Also, there is little overlap in these ranges. When a band is observed in one of these regions we can assign it to the presence in the molecule of that particular group and we can be fairly certain of the assignment. For example, the band at 1750 $cm^{-1}$ in Figure 3.14 must be due to the presence of a carbonyl (>C=O) group as it is the only one that absorbs in this region. If there is no absorption near here the sample is not a carbonyl-containing compound. Many biologically important molecules contain this group; proteins, fatty acids and triglycerides all show absorption bands between 1800 and 1700 $cm^{-1}$ although simple sugars and polysaccharides do not.

Simple sugars are classified as aldoses (derived from aldehydes) and ketoses (from ketones). They might therefore be expected to show a carbonyl absorption. The lack of the group is explained by the formation of ring structures. You can find out more about this in a biochemistry textbook.

In the rest of the spectrum (1400–600 $cm^{-1}$) there is much more overlap of the ranges and much less certainty as to the assignment of a particular band unless other information is available. However, the region is very useful in providing confirmatory evidence for assignments made from the 4000–1400 $cm^{-1}$ region. The bands at 1250 and 1050 $cm^{-1}$ in Figure 3.14 are examples of this. The molecular formula requires there to be two oxygen atoms in the structure and these two bands could verify that a C—O group is also present. In fact, it is very likely to be an ester, which contains —(C=O)—O—. Notice that this region can tell you which molecular groups are not present just as well as which ones are present. The absence of bands near 3050 and 750 $cm^{-1}$ suggests that the compound does not contain a benzene ring.

The structure of an ester group is

$$-\overset{\overset{\displaystyle O}{\|}}{C}-O-$$

and the benzene ring is an important structural type. If you are uncertain about this sort of detail for organic compounds you should consult a textbook such as Brown, W.H. (1987) *Introduction to Organic and Biochemistry*, Brookes/Cole, ISBN 0 534 07386 7.

Thus we can establish important and detailed information about the structure of a sample from its mid-IR spectrum. With practice, an IR spectroscopist readily recognizes the patterns generated and can very quickly reach conclusions. Of course, these days, spectroscopists have computers to help them, particularly in searching for correlations through large libraries of spectra which may be outside the spectroscopist's experience.

### 3.5.4 Identifying samples

The pattern of absorption bands in the 1400–600 $cm^{-1}$ region is very sensitive to small changes in molecular structure – so much so that the region has become known as the **fingerprint region** because no two compounds have been found to give identical spectra in this region. Thus another important use of the mid-IR spectrum is in the identification of test samples. If a complete match of peaks is found between the test and a standard spectrum, the test must be identical to the standard. There are tens of thousands of standard spectra now available, so you would not expect to search through them all, even using computer-searching techniques. But the search can be narrowed down from the diagnostic information obtained from the 4000–1400 $cm^{-1}$ region. You would also expect to use information from other types of analysis. Even so, there are millions of known organic compounds and you can only match standardized spectra. Thus the matching may well give only the best fit rather than a precise identification.

### 3.5.5 Determining impurities

We have described the above process in terms of the ideal situation where the samples are pure. Impurities may not show up clearly even though they may account for as much as 10% of the sample. However, there may well be a lack of clarity in the peaks as compared with the spectrum of the pure material, in which case you will suspect that the sample is impure. Occasionally one or more of the impurity peaks may be distinct from the rest and you may be able to form some idea as to what the impurity is. If the identity of the sample is known its pure spectrum can be subtracted from the impure version (using microprocessor software, of course) to reveal the spectrum of the impurity. If this is suspected to be just one compound it can be classified and even identified, using a computer search if necessary. Of course, the same reservations about identification apply as for pure samples.

### 3.5.6 Mid-infrared spectrophotometers

There are two main categories of these instruments used in analytical laboratories.

*Dispersive instruments*

These are so called because the radiation from the source is dispersed into the range of wavelengths, and the wavelength of interest is selected from the range by a system of slits, as we have encountered in UV–visible spectrophotometry (Section 3.3.1). Such instruments have an optical layout of the double-beam type (see Figure 3.16) and scanning is the most usual mode of operation.

There are, as you might expect, differences in detail from UV–visible instruments, the most significant being the materials used and the design of some of the components. The source is now a miniature equivalent of an electric fire (several versions are used, depending on the sophistication of the machine). The monochromator is placed after the sample compartment (to allow the maximum power through the sample) and the detector is most likely to be a thermosensitive integrated circuit, although there have been several other types, including the thermocouple, in wide use.

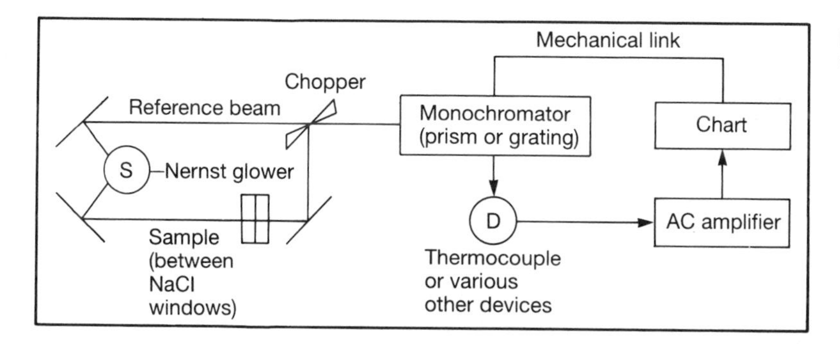


**Figure 3.16** Block diagram of a double-beam IR spectrophotometer.

*Fourier transform spectrophotometers*

The theory and specialized use of the FT technique have been available for a long time, but only in the last 15 years or so have computers allowed the technique to be applied to routine instrumentation. The basis of the system is an **interferometer**, which replaces the monochromator of the dispersive system. It produces an interference pattern from the light beam which has the same relation to the normal spectrum as a hologram does to a picture. However, in contrast to the regeneration of a picture from a hologram, FTIR interference patterns are transformed into normal spectra by the instrument's microprocessor. The machines are effectively single beam in their operation and so a background spectrum has to be recorded (or stored) first. In spite of the vast amount of calculating that is done, the production of spectra is significantly quicker than for a dispersive machine. However, for spectra of similar quality, the FT machines are rather more expensive.

## 3.5.7 Computer linking

FTIR machines have to have access to powerful computing facilities. This may be built into the machine or it may be provided by an attached computer. Most modern dispersive machines now provide similar facilities. Once the spectrum has been recorded it can be displayed and manipulated on screen in order to get as much information from the data as possible. You can expand regions of particular interest to see more detail, convert from percentage transmission to absorbance, which is useful in quantitative work, and compare two spectra or subtract one spectrum from another, which is helpful in identifying impurities.

## 3.5.8 Materials transparent to infrared radiation

Remember the way a greenhouse works is that the glass is transparent to some IR wavelengths but absorbs others.

The materials used for containing samples in UV–visible spectroscopy – plastic, glass and quartz – all absorb some parts of the

**Table 3.3** IR-transparent materials

| *Substance* | *Transparent range ($cm^{-1}$)* | *Comments* |
|---|---|---|
| NaCl | beyond 4000–600 | Most common, general purpose, sensitive to moisture |
| KBr | beyond 4000–400 | Often used for solid samples (as in KBr discs), sensitive to moisture |
| CsI | beyond 4000–200 | For work in the far-IR, sensitive to moisture, soft, easily scratched |
| AgCl | 10 000–400 | Insensitive to moisture, but very fragile and darkens on prolonged exposure to light. Soft, easily scratched |
| ZnS | 50 000–770 | Insensitive to moisture but not suitable for low-wavenumber work |

From Graseby-Specac Sampling Techniques booklet.

In the early days of IR spectroscopy the monochromators contained prisms made of salt.

IR spectrum. This makes them unsuitable for IR spectroscopy and alternative materials have to be found. The most widely used is sodium chloride (common salt!), which does not show any absorbance until about 600 $cm^{-1}$. For work beyond this (which is effectively into the far-IR) other similar materials may be used: potassium bromide (to 400 $cm^{-1}$) and caesium iodide (to 200 $cm^{-1}$). You normally expect these substances to be granular solids but for this work they are formed into clear transparent discs about 2 cm in diameter. They are rather fragile, easily scratched and very susceptible to the effects of moisture from the air or from samples. Thus they are kept in a desiccator and repolished when their surfaces have become cloudy or uneven. Other materials are available for specialist work (see Table 3.3).

### 3.5.9 The reference beam

Remember the greenhouse effect. It is due to carbon dioxide and other gases absorbing IR radiation.

Two substances present in the air absorb IR quite strongly: carbon dioxide and water vapour. In IR work we compensate for this by using a double-beam system and passing the reference beam through air, but, in contrast to UV–visible practice we do not usually have a cell in this beam (it is generally too difficult to make two cells similar enough to obtain satisfactory compensation). The result is that the baseline is not at 100% $T$ but at around 80–90% $T$ (see Figure 3.14).

### 3.5.10 Sample-handling procedure

These days you may be able to correct for the thickness by changing the settings on the machine but this can lead to distortions of the spectrum.

**For neat liquids** (not solutions) you put a drop or two of the sample on the face of one salt disc and then place another disc on top of it, making sure that no air is trapped in the thin film sandwiched between them. Then you place the pair of discs in a metal holder and insert the assembly into the machine. The transmittance at 4000 $cm^{-1}$ should be about 70–80%. If it is less than this the sample is probably too thick and may have to be rerun with a smaller amount. It is also possible to have a sample too thin, in which case the strongest peaks reach only to about 40% *T*. You should rerun such a sample with a larger amount.

There is too much scattering of the radiation from the neat solid, but the presence of the oil reduces it and gives better transmission through the sample. You must, of course, allow for the absorption of the Nujol itself in the spectrum obtained. It gives bands at 2900, 1460 and 1380 $cm^{-1}$.

**For solids** either (a) grind a small amount of the sample (1–2 mg) with about three drops of **Nujol** (a sort of paraffin oil) to give a milky 'mull' and then continue as for liquids, or (b) grind a small amount (1–2 mg) with about 100 mg of dry potassium bromide. You then apply a very high pressure to the mixture in a die to produce a thin translucent **KBr disc**. Place this in a special holder and insert the holder into the machine.

If the grinding is not sufficient, you will get a range of particle sizes in the sample and this produces a sloping baseline in the spectrum, rising from a low %*T* at 4000 $cm^{-1}$ to the normal level by about 2000 $cm^{-1}$.

**For gases** longer path-length cells are required. You can use a glass cylinder 5–10 cm long for reasonably dense gases, with salt end-windows of course. Taps at the ends of the cylinder allow for filling and emptying the cell. Trace amounts of gas, including low levels as might be found in environmental studies, need much longer path-lengths, up to 20 m. These are achieved by having a mirror system within the cell so that the radiation is reflected several times back and forth before it emerges and is detected.

At 'high resolution' (equivalent to a high magnification of the wavelength scale) gas spectra can be seen to be composed of many very narrow absorption bands (effectively lines). These are due to the rotation of the molecules. In solid or liquid spectra the lines become merged into the smooth shape of normal bands.

**For solutions** the solvent has to be chosen with care because it has its own absorption bands and these must not obscure too much of the spectrum, especially in the regions of principal interest. For a cell thickness of 0.1 mm, commonly used solvents are **tetrachloromethane** (carbon tetrachloride) which obscures the range 820–720 $cm^{-1}$ and **trichloromethane** (choloroform) which obscures the ranges 3020–3000, 1240–1200 and 805–600 $cm^{-1}$. Water is rarely used as a solvent in IR work since it obscures a large part of the spectrum, and dissolves the most common IR-transparent materials. Insoluble materials are much more expensive and have practical drawbacks (see Table 3.3).

A particular difficulty with solutions is that you have to use a reference cell to compensate for the absorption of the solvent. However, it is difficult to get two cells to match closely because of the short path-lengths. You can get variable path-length cells so that you can adjust their path-lengths to match each other but they are expensive.

**Special cells** are available in a wide variety of styles. For example, opaque samples can be measured by reflectance (compare NIR, Section 3.4); micro-cells are used for very small samples.

**Common applications of IR**, as we have described above, are:

- determination of functional groups within a molecule;
- identification of molecules;
- determination of impurities in samples.

**Figure 3.17** Exploded diagrams of typical IR cells. (a) A simple holder for neat liquids and for 'mulls' (a paste of solid ground up with a carrier liquid). The spacer may be omitted. (Reproduced from B. George and P. McIntyre, *Infrared Spectroscopy*, published by Wiley, 1987.) (b) A cell for solutions. The spacer usually has a precise thickness in the range 0.1–1 mm. The solution is introduced after the cell has been assembled. (Courtesy Specac, UK.) In both cases the windows are made from sodium chloride or a similar material. (c) A typical infrared cell for gas samples.

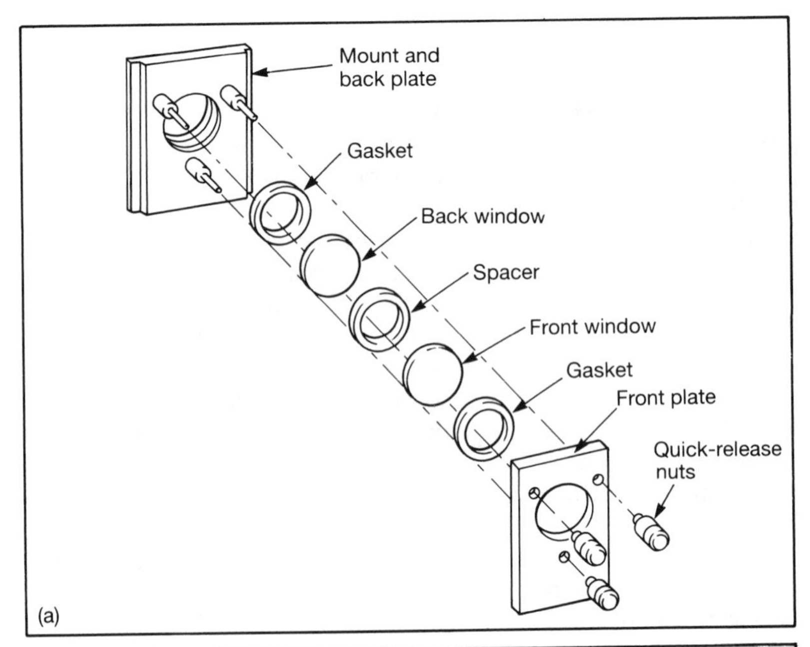


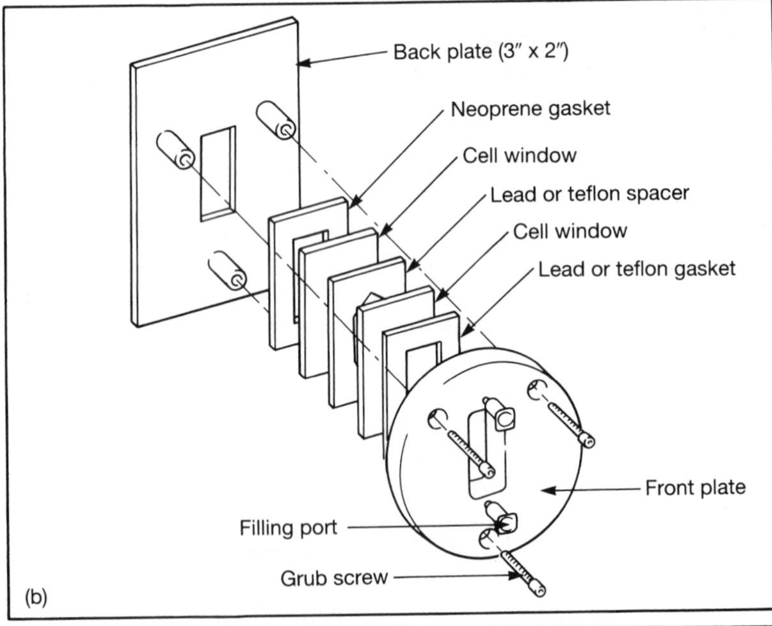


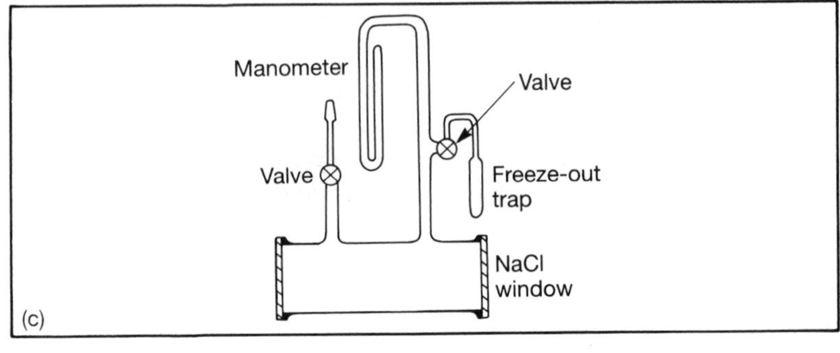

**Less common applications** include the following.

- Determination of unsaturation in margarine; a comparison has been made between the intensities of the bands at 3030 $cm^{-1}$ in butter and margarine. The band arises from the CH=CH groups and is more intense in the margarine sample. The procedure is much more rapid than gas chromatography.
- Determination of oil in water; since oil is almost insoluble in water the levels are very low. A special cell is needed because of this. Fortunately the absorption of the oil does not occur at wavelengths where the water absorbs strongly.

## 3.6 Fluorescence spectroscopy

We have already mentioned some examples of fluorescence in everyday life and described the molecular processes in their origin (Section 3.2). Laboratory uses of fluorescence include the fluorescent staining of thin sections of plant or animal material for microscopic work. Using UV illumination particular structures in cells show up clearly even though they might be obscured by other structures in visible light microscopy. Also, bands of material in DNA and protein chromatography or electrophoresis can be made to show up with a greater sensitivity than with normal dyes. Now we will look at the analytical applications and the instruments used in fluorescence spectroscopy. However, it would be useful to consider first the advantages and disadvantages of the technique.

### 3.6.1 Advantages of fluorescence spectroscopy

It is **highly sensitive**. This is because the system detects emitted light and very low levels can be amplified reliably.

More properly we should say that it has very low detection limits.

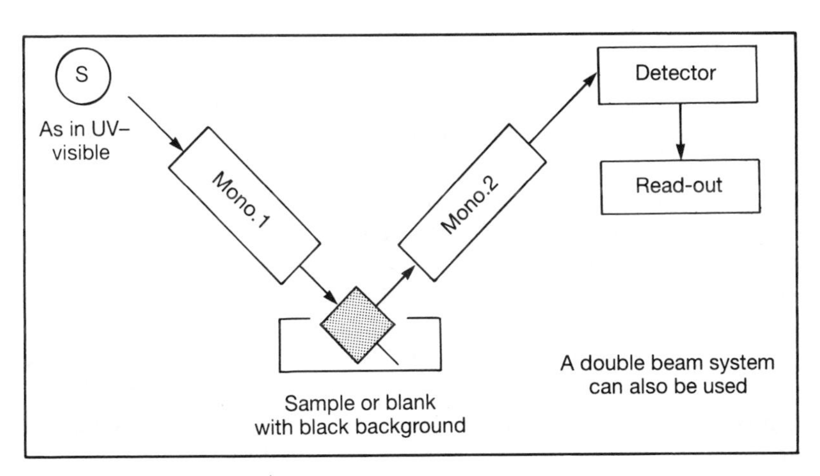


**Figure 3.18** Fluorescence spectrophotometer layout.

It uses **simple instrumentation** for routine machines (see Figure 3.18) which give acceptable precision although, of course, more sophisticated machines are available for advanced applications and research.

It is a **selective technique**. Not all compounds fluoresce and two wavelengths have to be specified in setting up the system (see below). This enables measurements to be done even though there may be other components in the sample.

### 3.6.2 Disadvantages of fluorescence spectroscopy

Because **not all compounds fluoresce** (and there is no theory for the reliable prediction of fluorescence) you cannot be sure that the substance you want to analyse can be done by the method until you have tried it! Compounds that do not fluoresce can be labelled with a fluorescent molecular group and then analysed. The disadvantage is that it requires **extra work** and you have to be sure that the labelling process does accurately reflect the levels of the component in the sample.

The fluorescence process is susceptible to **quenching**, which makes the emitted intensity less than predicted, sometimes to such an extent that the fluorescence is virtually extinguished. It can be caused by two main sorts of interference: **chemical**, where the presence of some substances such as molecular oxygen or some acids interferes with the transfer of energy from the excited molecules; and **colour**, when substances in the solution absorb the emitted light (it may even be the fluorescent substance itself!).

The exposure of the sample to the high energy of UV radiation can lead to **sample decomposition** while the measurements are being made. However, modern machines allow for this by blocking the light path to the sample except at the time it is being measured.

### 3.6.3 Excitation and emission spectra

For the best performance in fluorescence spectroscopy we need to know the wavelength which produces the most excitation and the wavelength at which the emission is most intense. As in absorption spectroscopy, where we did a scan to find the value of $\lambda_{max}$, here we scan first to find the wavelength of the incident radiation which produces the most intense total emission, and then, having set the excitation to this wavelength, we scan through the emission wavelengths to find their maximum. The results can be recorded graphically by most machines and a typical result is given in Figure 3.19. Notice that, as was stated in Section 3.2, the emission is at longer wavelengths than the excitation. Excitation requires the incident radiation to be absorbed by the sample, and

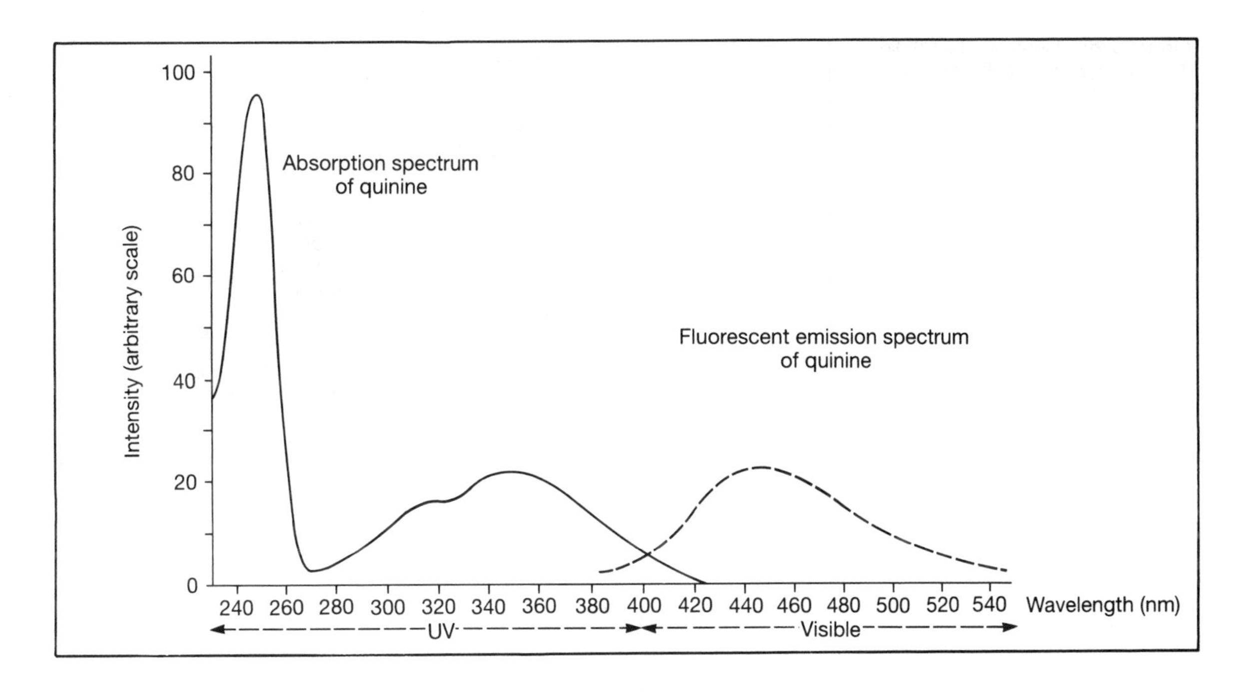


**Figure 3.19** Absorption and fluorescence spectrum of quinine. For the most sensitive and selective quantitative analysis of quinine you would use an excitation wavelength of 245 nm ($\lambda_{max}$ of the more intense absorption band) and measure the emission at 445 nm where it is most intense.

it is not surprising that the excitation spectrum is very similar to the absorption spectrum. Also, because of similar arrangement of vibrational energy levels in both the ground state and the first excited state, the excitation and the emission spectra are often rough mirror images.

### 3.6.4 Fluorimeters

Check Figure 3.18 (the block diagram of a fluorimeter) and notice that several of the components are the same as for a single-beam absorption spectrophotometer (Figure 3.7), namely the source, the detector and the amplifier system. The main differences are in the use of two monochromators (one for excitation and one for emission) and the optical arrangement at the cuvette (the right-angle reduces the amount of scattered light from the incident beam reaching the detector). The cuvette itself has to be made of quartz since almost invariably the exciting wavelength is in the UV range, and the right-angle arrangement requires that two adjacent faces are clear – in practice, all four are (in contrast to absorption cuvettes, as described in Section 3.3.3).

### 3.6.5 Procedure

The procedure for operating a fluorimeter is illustrated in Figure 3.20.

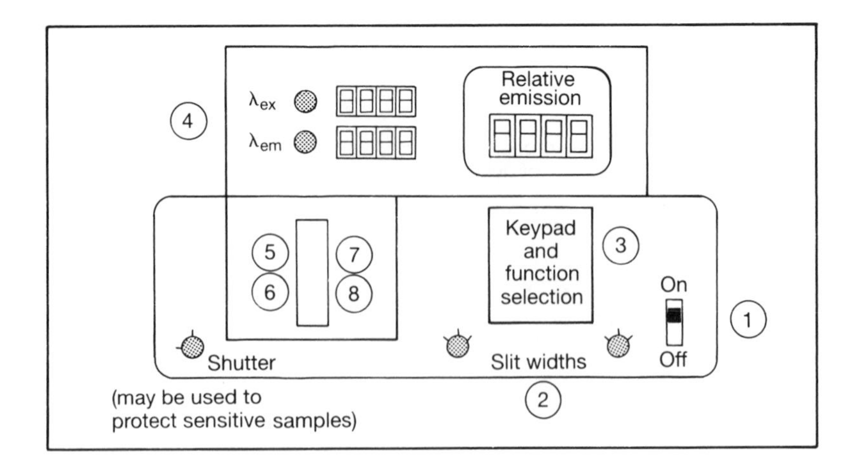


**Figure 3.20** Operation of a simple fluorescence spectrophotometer.

### Initial settings

1. Switch on and allow time for the instrument to warm up and stabilize. This may be incorporated into a self-checking routine if the machine is microprocessor controlled.
2. Set the slit widths to appropriate values (if this facility is available). If you do not have any information about this, use the settings regarded as general purpose by the manufacturer.
3. If necessary perform scans to establish the best wavelengths for excitation ($\lambda_{ex}$) and emission ($\lambda_{em}$). Most modern machines have facilities for doing this automatically as well as manually.
4. Set the monochromators to these wavelengths. Older (or cheaper) machines may use special filter glasses to do this, but since these select a range of wavelengths you may get less precise results.

### Calibration

5. Insert a suitable blank and adjust the reading to zero. If you found a high reading just before you do this you should check carefully that there is no contamination in the sample.
6. From your series of standards insert the highest concentration and set the instrument to a convenient figure. The scale measures the relative emission and so has an arbitrary range of values. Often a range of 0–100 is satisfactory, but if you have already checked the linearity (step 7) you could use a scale which gave the concentration reading directly.
7. Starting with the lowest concentration standard, record each in turn and check the linearity of the calibration (by a graph, or with a calculator or computer). A straight line through the origin or a slight curve is acceptable. A distinct curve, which might even show a maximum, indicates quenching and will

not be reliable. In such a case the range of concentrations of the standards must be lowered and steps 3–7 repeated.

**Measuring test samples**

8. Insert test samples as necessary and take readings. If any are greater than the maximum for the standards, they must be diluted by a known factor and re-read (see Section 1.6). It is worth doing preliminary rough calculations to try to ensure the test samples give readings within the range and avoid this extra work. It is very important in fluorescence that the standards are made up in a way which is closely similar to the test samples.
9. The concentrations are determined from the calibration graph (or by means of a microprocessor-based system) in the normal way (see Section 1.6).

### 3.6.6 Applications

Quinine fluoresces if it is in an acidic solution. It is present in some medicinal tablets as well as in tonic water and can readily be detected and measured by fluorimetry.

Many enzymes catalyse redox reactions (see Section 4.1), and many of them use $NAD^+$/NADH as a coenzyme. The NADH fluoresces and so it can be used to follow the progress of such enzyme reactions. This can also be used to determine the concentration of the substrate of the enzyme.

Proteins and amino acids can be reacted with *o*-phthalaldehyde to give a highly fluorescent product and so their concentrations can be determined with high precision.

DNA fragments bind ethidium bromide to give a fluorescent complex. This enables them to be detected in electrophoresis gels (Section 6.2) and their quantities determined.

## 3.7 Atomic spectroscopy

In atomic spectroscopy we shall be dealing with methods which enable us to analyse which **elements** are present in a sample, in contrast to determining which molecules it contains, and, more importantly, we can determine the levels of these elements. The term 'levels' is not specific enough for the actual analytical work but is used here to indicate such quantities as $mg\,dm^{-3}$ for solutions or $mg\,g^{-1}$ when the original samples are solids.

In any analytical technique we need to be aware that the signal we measure may be affected by neighbouring atoms and molecules and, for atomic spectroscopy, we must have **free atoms** so that there is no interference from bonding to other atoms or

Why might we expect free atoms to be rare? Remember that most atoms have to form bonds in order to produce stable electronic configurations. Are there any substances that consist of or contain free atoms? The noble gases do and mercury vapour but most other atoms are too reactive.

from interactions with molecules. However, free atoms are rare in normal conditions so we have to use some method of breaking down molecules into free atoms. A flame is the most widely used method, but others are available such as a very high-temperature graphite rod, a plasma generated by a radio frequency field or in some cases the decomposition of the hydride of the element.

Once we have free atoms we can produce atomic spectra which consist of lines rather than bands (see Section 3.2 and Plates 2 to 5). This allows us to be quite selective even when there are several different elements in the sample. The chances of their spectral lines overlapping are much less than with bands, and we can almost certainly find a particular line that is clear of all others. Also, we can use either emission or absorption. The reasons why we might choose one rather than the other are given below.

### 3.7.1 Flame emission spectroscopy (flame photometry)

The easiest method of detecting the emission of light is by eye! You may be familiar with the flame tests for some elements that form part of most introductory chemistry courses. When a small amount of a compound of one of these elements is introduced into a colourless flame a distinctive colour is produced which can help to identify the element. If the light from the flame is examined in a spectroscope it is seen to be a set of lines of particular wavelengths which are unique to that element. Elements which show this phenomenon most clearly are the metals in groups 1 and 2 of the periodic table, and a few others. A list is given in Table 3.4. It is not a long list but we have only considered the emission of visible light. We should also be asking if some elements emit UV or IR radiation in similar circumstances. Theoretically many do, but in fact flames are not really hot enough

**Table 3.4** Flame colours and diagnostic wavelengths of some elements

| *Element* | *Flame colour* | *Wavelength (nm)* | *Comments* |
|---|---|---|---|
| Sodium | Yellow | 589 | Familiar as the colour of street lights<br>Also known as the 'sodium D-line'<br>High resolution shows it to be actually two lines |
| Potassium | Lilac | 766 | |
| Lithium | Bright red | 670 | |
| Strontium | Brick red | 660 | Not easy to distinguish from lithium |
| Barium | Green | 553 | |
| Copper | Blue | 500 | |

to produce UV emission and IR emission is swamped by the flame itself.

The process of converting molecules into free atoms is known as atomization and it reqires energy to break the bonds. Emission requires excited atoms and needs an input of energy considerably greater than that required for atomization. The excitation energy varies from one element to another and this affects the proportion of atoms that become excited. A higher excitation energy means fewer excited atoms. The proportion also varies with the temperature, a higher temperature giving a greater number of excited atoms. However, it is only the elements mentioned in Table 3.4 that have a low enough energy of excitation to show any significant emission at the temperatures of most commonly available flames. The proportion of excited atoms in the flame for such elements may be up to about 10% but for other elements it is usually much less than 1%. Thus the emission technique is inherently insensitive, and produces detection limits which are not low enough for much modern analytical work. Nevertheless flame emission requires simple apparatus, is easy to use and produces results which are quite adequate for some routine analyses such as sodium and potassium levels in blood and other biological fluids, or in water samples that may be subject to environmental investigation.

When the concentration in the flame is low, the intensity of the emission is proportional to that concentration and hence to the concentration in the sample. At higher concentrations self-absorption becomes noticeable, in which some of the light emitted is reabsorbed if it has to travel through the flame to reach the detector. The atoms responsible for the absorption are the unexcited atoms of the **same** element and, from what we said in the previous paragraph, we know there is always a large proportion

The reabsorption of emitted light must take place at the same wavelength because the energies must be the same. This only happens in atomic spectra when the same energy levels in the same element are involved.

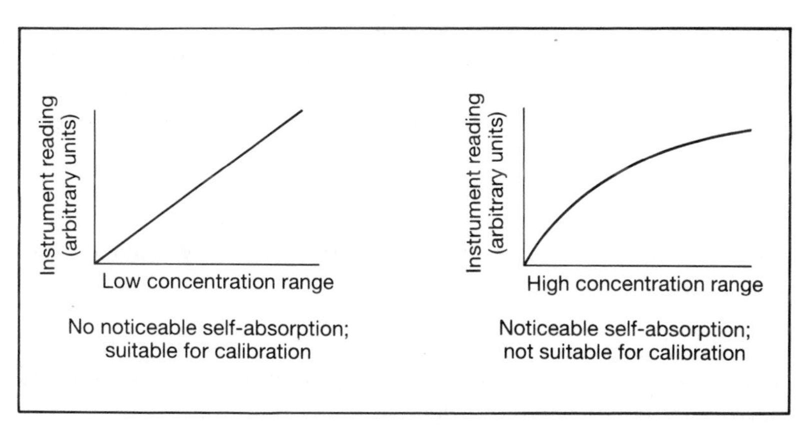


**Figure 3.21** Flame photometry calibration graphs.

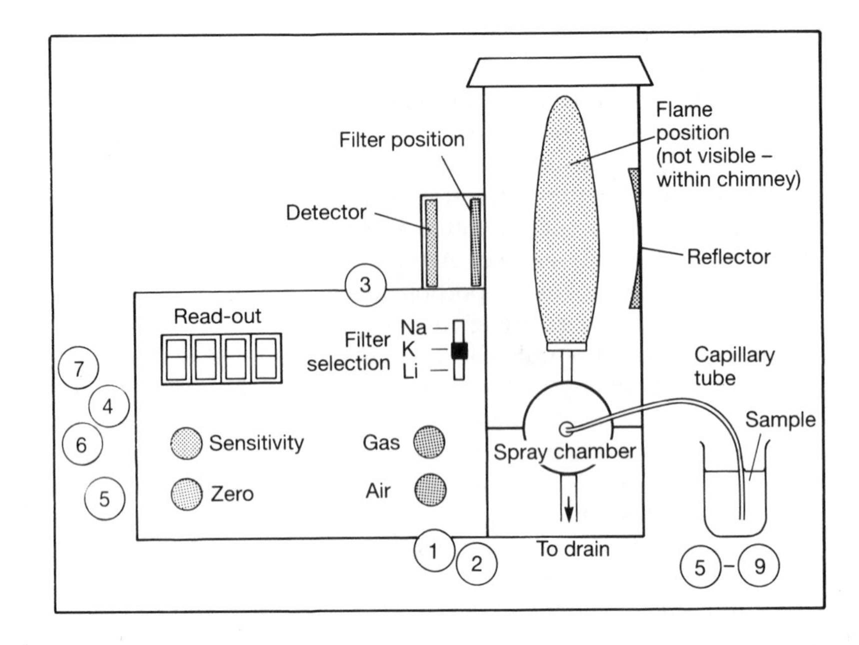


**Figure 3.22** Operation of a typical flame photometer.

of such atoms in the flame. Self-absorption produces a negative deviation from linearity in the calibration graph. If the curvature is not too severe the graph can still be used, but for greater reliability you should use more dilute standards in the calibration. The curvature is then virtually absent.

*Instrumentation*

The components of a flame photometer are shown in Figure 3.22.

*Spray chamber.* The sample solution is sucked into the flame by the venturi action of the nozzle in the chamber. Only the finest droplets are small enough to evaporate, vaporize and become atomized in the time it takes for them to rise about 4 cm in the flame. A baffle system in the spray chamber prevents the passage of larger droplets, which drain out at the bottom of the chamber.

*Flame area.* The flame size and temperature can be adjusted by controlling the flow rates of the fuel gas and of the air. Natural gas (methane), propane or butane are commonly used as fuel gases. The air may be pumped in by a compressor.

*Wavelength selection.* Coloured glass filters are normally used since the spectral lines in the visible range are well spaced and narrow-range wavelength selection is not required.

*Detector.* A simple photocell is sufficient to give an adequate response throughout the visible range.

*Read-out.* The read-out may be a meter but is most likely to be a digital display. It shows intensity of emission using an arbitrary scale. You can adjust its sensitivity and it may be possible to make it give a direct read-out of concentration, but such a calibration will only be reliable if the response has been found to be linear.

### *Operating procedure*

**Preliminaries**

Make sure your supply of deionized (or distilled) water is fresh and free from contamination. All plastic and glassware should be rinsed appropriately before use with the next solution that it is to contain.

1. Switch on and **start up** gas supplies; ignite the flame. If you have not done this before, you will need help.
2. **Optimize** the flame setting. Again, you may need help.

**Calibration of the instrument**

3. **Select** the appropriate **filter**.
4. Set the **sensitivity** to approximately mid-way in its range.
5. Put fresh deionized water into a rinsed beaker and insert the suction tube into it; **set the zero** of the instrument. You will probably not get an absolutely steady reading because of flickering in the flame, but the reading should be close to zero all the time and there should be no visible colour other than the very pale blue of a normal gas flame.
6. Put the highest concentration you wish to use into a beaker and allow it to be sucked into the flame; **set the upper reading** (using the **sensitivity** control, not the zero control). This can be any sensible value since it is an arbitrary emission scale, but 100 is often chosen. Again, the reading will not be absolutely steady, but it should not be varying much and should not be drifting.
7. Return to deionized water and **recheck the zero** (using the zero control, not the sensitivity). It is good practice to do this at intervals throughout the analysis, but it is not necessary between every sample, as long as you leave long enough for the new sample to flush out the previous one.
8. Take readings of each of your **standards**, without altering any

of the controls, and check the linearity of the calibration (by graph, calculator or computer). As we mentioned for Figure 3.21, a straight line or slight curve is acceptable. A distinct curve requires lower concentrations for your standards, in which case you should prepare new ones and repeat steps 5–8.

**Measuring test samples**

9. Without changing the controls further, take readings for your test samples and use the calibration to determine their concentration. Because of the slight variations in the flame, take the average of several observations. As always, if the reading is beyond the calibrated range the test sample must be diluted to bring it into the reliable range.

*Applications of flame photometry*

Sodium and potassium levels in blood serum are important in medical diagnosis. Diluted samples of serum can readily be measured for these elements. It is one of the commonest tests done in clinical laboratories.

Lithium is used in the treatment of some mental conditions. Flame photometry can be used to check that the dose in a pharmaceutical preparation is correct.

Calcium is a particularly important mineral in baby foods. The food would be treated with nitric acid to release the calcium into solution, which can then be measured by this technique.

### 3.7.2 Atomic absorption spectrophotometry

In discussing flame emission spectroscopy we pointed out that many elements produce a very low proportion of excited atoms in a flame generated from natural gas and air. Other gas mixtures give higher temperatures, as do alternative atomizers such as the graphite-rod furnace, but they still do not produce sufficient excitation to make the emission process feasible as an analytical technique. On the other hand, the proportion of unexcited atoms in flames is very high (greater than 99.5%) and, as with the self-absorption in flame photometry, these atoms can absorb light of the wavelength that the excited atoms emit. Thus, for atomic absorption, we shine light of the correct wavelength into the flame and measure how much its intensity is diminished as it passes through the flame. Since this is an absorption process Beer's law applies. We will now be working principally in the UV region although some elements are measured in the visible region, including those mentioned in flame photometry.

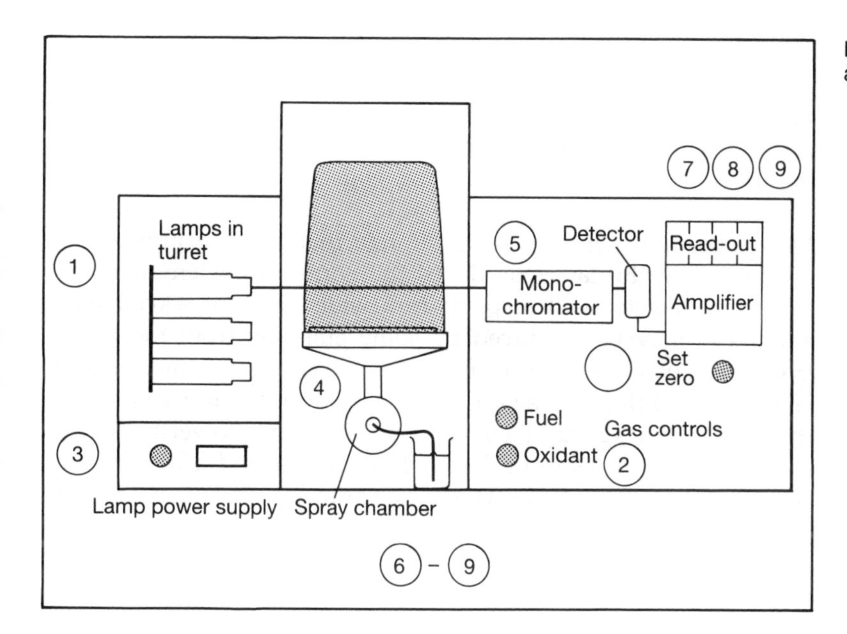


**Figure 3.23** A typical atomic absorption spectrophotometer.

*Instrumentation*

In some respects the components of an atomic absorption spectrophotometer are more elaborate versions of those in a flame photometer. However, there are important differences and additions; Figure 3.23 shows a typical instrument.

*Light source.* This has to be a special lamp which produces exactly the right wavelength since even the best monochromator cannot produce the narrow bandwidth (see Box 3.5) required for precision work. Such a lamp is known as a **hollow cathode lamp** and its cathode contains some of the element that is to be analysed. In the lamp these atoms are excited and so produce the wavelengths that the unexcited atoms in the flame will absorb. You may notice from this that one of these lamps can only be used for one element. Some instruments incorporate a turret holding several lamps for ease of changing to other elements, but this cannot be done while the sample is running through the flame. There are a few lamps which contain more than one element. They are, of course, much more expensive.

*The spray chamber.* This serves the same purpose as in flame photometry and functions in a similar manner.

*The burner.* This has a long thin slit so that the light path through the flame is as long as possible and can give maximum sensitivity. The position of the burner is adjustable so that you can make sure you have the best alignment of the beam and the flame. Different parts of the flame produce different temperatures and either oxidizing or reducing conditions. The height of the beam above the slit of the burner is chosen to best suit the requirements of the element being analysed. For very small samples an alternative to the flame, the graphite-rod furnace, is used.

*Monochromator.* It may seem strange that a monochromator is needed since the lamp produces the correct wavelength. However, the detector will respond to a wide range of wavelengths and these may be produced by some emission from the flame, from molecular species (not fully atomized) in the flame, or from laboratory lighting and so on. A good monochromator will reduce this sort of background effect to a minimum. Nevertheless, in some cases the background is still significant and a background correction has to be made. There are various ways of tackling this; some of them are very sophisticated and involve expensive additions to the system. You would have to consult the instrument manual and specialist texts if you needed to pursue this further.

*Detector.* As with UV spectrophotometers, this is a photomultiplier tube.

*Read-out.* This should give values of absorbance.

*Operating procedure (refer to Figure 3.23)*

An atomic absorption spectrophotometer is more sophisticated than a flame photometer so **you will need help at first**. Some gases such as hydrogen and acetylene (ethene) need particular care because of their flammability and explosion risk and you should not handle their cylinders before being given proper instruction.

You are very likely to be determining low levels of metals and so it is very important that your laboratory ware is scrupulously clean and your solutions are prepared from the highest-purity materials. It is quite possible that your solutions can leach traces of elements such as lead from glass containers so plastic ones should be used. Conversely, extremely low levels of some metals can adhere to glass and thus not be detected.

**Preliminaries**

1. Make sure that the **correct lamp** is in position.
2. Switch on main **power**, set up the gas supplies and **ignite** the flame. You will need to know which gases to use for the particular element you are analysing. Charts and booklets giving you the information should be available with the instrument. A few examples are given in Table 3.5.

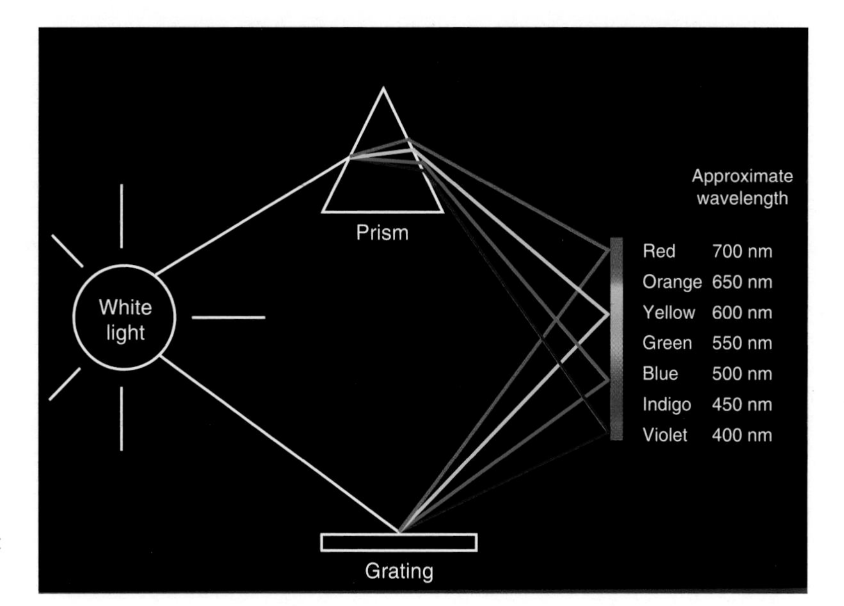


**Plate 1.** The dispersion of white light.
You can see spectra generated by the prism mechanism when you look at cut-glass objects and, of course, a rainbow. The diffraction grating effect is shown by compact discs and ordinary long-playing records.

**Plate 2.** The spectra of white light (top) and sunlight (bottom).
The spectrum of white light from, for example, a slide-projector or a car headlight is a continuous range of colours from violet to red. In the sunlight spectrum there are absorption lines, known as Fraunhofer lines, superimposed on the white light spectrum because the white light, generated in the lower layers of the sun's atmosphere, passes through the outer layers which contain the atoms of many elements.

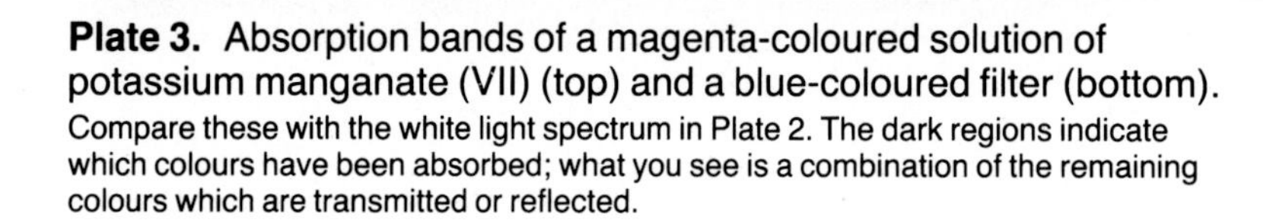

**Plate 3.** Absorption bands of a magenta-coloured solution of potassium manganate (VII) (top) and a blue-coloured filter (bottom).

Compare these with the white light spectrum in Plate 2. The dark regions indicate which colours have been absorbed; what you see is a combination of the remaining colours which are transmitted or reflected.

**Plate 4.** The emission spectrum of neon.

You can easily see what we mean by a 'line spectrum' from this plate. The predominant intensity is in the red region which is the colour we see when we look at a neon sign. Other gases are used for other colours.

**Plate 5.** The emission (top) and absorption (bottom) spectra of sodium atoms.

Sodium street lights give the yellow colour shown in the top spectrum. The main emission from excited atoms of sodium is two very closely spaced wavelengths (589.0 and 589.6 nm) which are not fully resolved in this picture. There are other faint lines in the visible spectrum of sodium and there are other lines in the UV region. The wavelengths absorbed by non-excited atoms of sodium are the same as the wavelengths in the emission spectrum.

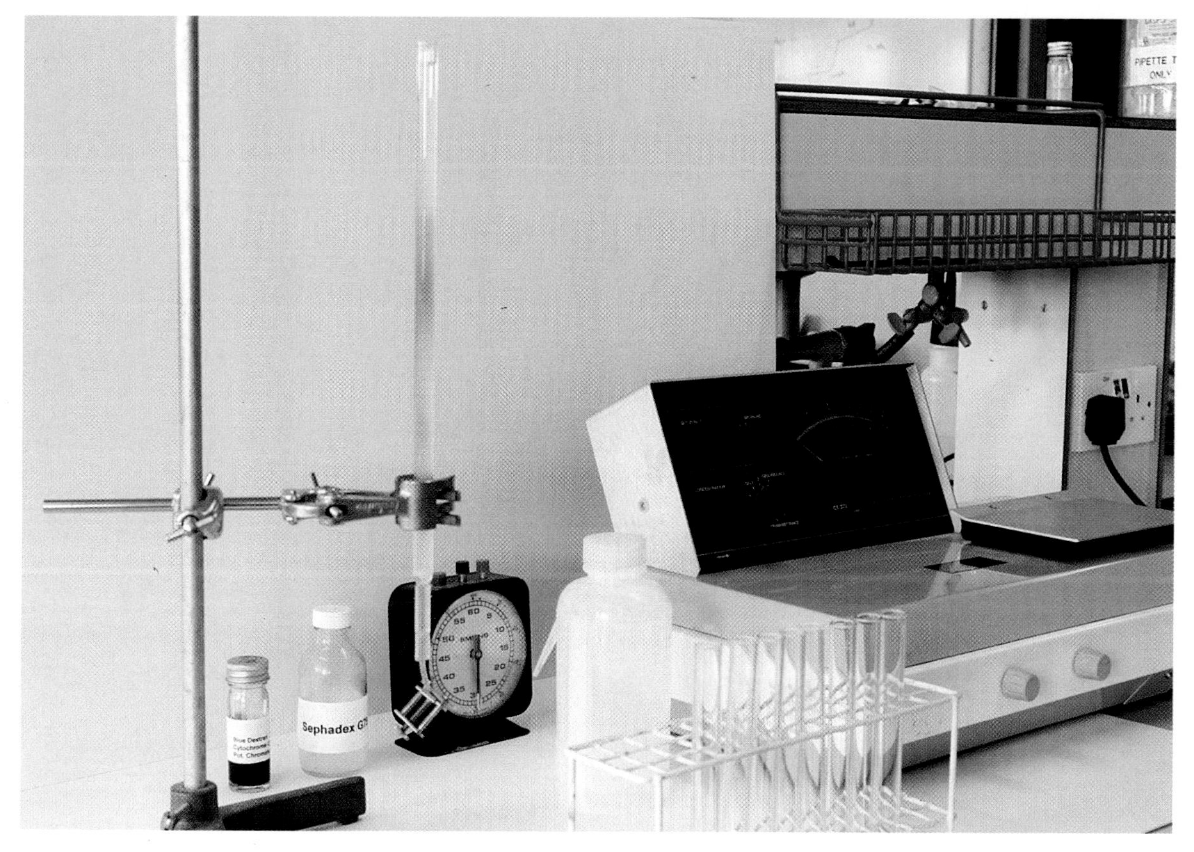


**Plate 6.** A typical simple chromatography column.

This is a demonstration of gel permeation chromatography in which large molecules (lowest band: blue dextran, $M_r$ 2 000 000 approx.) are separated from smaller ones (middle band: cytochrome C, $M_r$ 12 500 approx.) which in turn are separated from the smallest which may be ions from salts (uppermost band: chromate ions, $M_r$ 116). Coloured molecules are used here to show the separation process. The separated components can be collected in the test-tubes and their concentrations determined in the spectrophotometer shown. Much chromatography is done with colourless substances, in which case the detection may be done using UV absorption. Often the separations are produced by adsorption effects rather than molecular size but the production of zones containing the different components is the same.

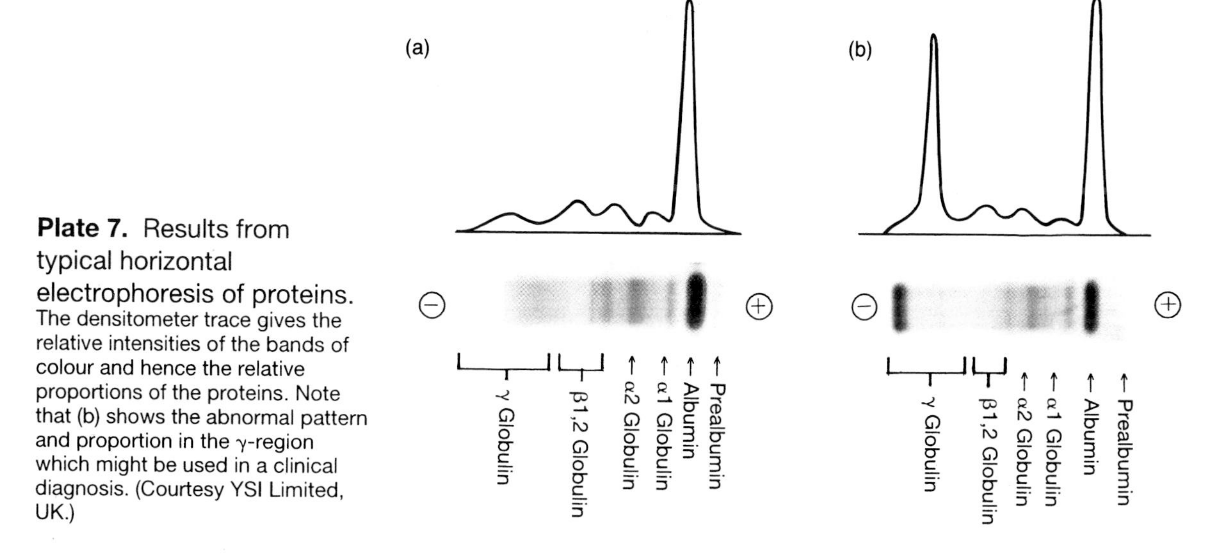


**Plate 7.** Results from typical horizontal electrophoresis of proteins.

The densitometer trace gives the relative intensities of the bands of colour and hence the relative proportions of the proteins. Note that (b) shows the abnormal pattern and proportion in the γ-region which might be used in a clinical diagnosis. (Courtesy YSI Limited, UK.)

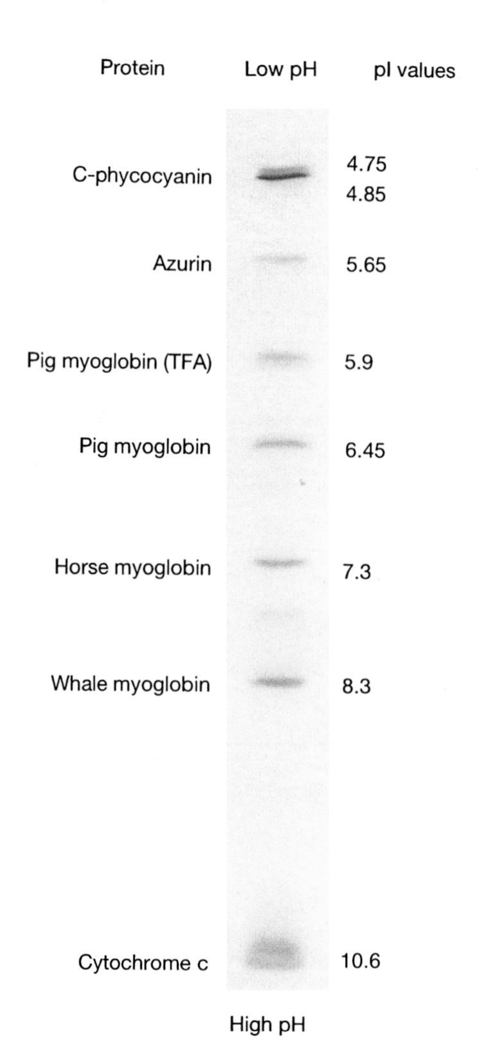


**Plate 8.** Isoelectric focusing of a mixture of coloured proteins.
These proteins are used to calibrate an isoelectric focusing experiment. Their isoelectric points are known and their positions can be clearly seen so values for unknown components of mixtures can readily be determined. You can see the sharpness of the bands produced by the focusing process.

**Table 3.5** Atomic absorption flames and interferences

| *Element* | *Wavelength (nm)* | *Fuel/oxidant* | *Interference* | *Remedy* |
|---|---|---|---|---|
| Barium | 553.6 | $C_2H_2/N_2O$ | CaO emission | Use background correction |
| Calcium | 442.7 | $C_2H_2/N_2O$ | Al, Fe, $SO_4^{2-}$ | Sr, La act as releasing agents |
| Copper | 324.7 | $C_2H_2$/air | None | |
| Sodium | 589.0 | $C_2H_6$/air | Ionization | K, Cs suppress the ionization of sodium |
| Lead | 217.0 | $C_2H_2$/air | None | |
| Zinc | 213.9 | $C_2H_2$/air | Background emission | Use some form of background correction |

3. Switch on the **lamp** circuit and bring to the recommended power.
4. Check the **alignment** of the burner, if required. Modern machines may be pre-aligned.
5. Set the monochromator. There may be a special procedure for this in case the scale is not calibrated precisely enough.

**Calibration**

6. Establish **zero absorbance** with high-purity deionized water.
7. **Either** take readings (average of three) for each of the **standards** in your range and check the linearity of your calibration (graph, calculator or computer). As with other spectroscopic techniques, a straight line or a slight curve is acceptable. A distinct curve may indicate **interferences** (see Table 3.5).

   **Or** use standard additions (see step 9).

**Test samples**

8. **Either** take readings (average of three), and determine the concentrations from your calibration graph.
   **Or**, when complex samples are used you may get greater accuracy by the method of standard additions (see Section 1.6).
9. **Standard additions**: measure the sample; add a known concentration of the standard and remeasure; repeat this twice; draw a graph as shown in Figure 1.7.

**Interferences** can arise from a variety of causes such as molecular fragments not fully atomized in the flame or ionization of some atoms, and various means are employed to overcome them as summarized for a few cases in Table 3.5. However, detailed discussion of this takes us beyond the scope of this book.

*Applications of atomic absorption spectrophotometry*

Tap water contains variable amounts of copper from the pipes and the brass fittings. It is important to be able to monitor the level in water supplies and it is very simply done using atomic absorption.

Much has been said about the danger of lead pollution, whether it is from car exhaust or other sources. Its level in the blood of children is best measured using the graphite furnace technique since it requires small samples. If you are looking for levels in tissues – animal or vegetable – you would probably digest the tissue with nitric or perchloric acid to remove the organic materials first.

Cadmium is a heavy metal that can be in wastes from industry and may consequently appear in estuary mud. It then enters the food chain of sea creatures and birds. Digestion of the mud with acid or alkali may be used to produce a solution suitable for analysis by this technique.

Some soils are deficient in essential nutrients for plants. They can be monitored in a similar manner to that for the estuary mud.

*Developments*

Inductively coupled plasma emission spectroscopy (ICP) is becoming widely used. It uses the extremely high temperatures that can be generated in low-pressure gases by radio-frequency coils. This enables most elements to be energized sufficiently for emission to be measurable, in contrast to the few that can be measured by flame photometry. Also it suffers from fewer interferences than atomic absorption. It is, however, a very expensive instrument to buy.

## Further reading

Harris, D.C. (1987) *Quantitative Chemical Analysis*, 2nd edn, W.H. Freeman, ISBN 0 716 71817 0.
Basic theory and practice of spectroscopy are given in Chapters 20–22, with some detail of the mathematics in an approachable style.

Braun, R.D. (1987) *Introduction to Instrumental Analysis*, McGraw-Hill International Edition, ISBN 0 07 100147 6.
Chapters 5–12
and
Skoog, D.A. and West, D.M. (1980) *Principles of Instrumental*

*Analysis*, 2nd edn, Holt-Saunders International Edition, ISBN 4 8337 0003 4.
Chapters 4–12.

These two books cover various types of spectroscopy and give detail of the components of the instruments.

The ACOL 'teach yourself' style books include the following titles.

Denney, R.C. and Sinclair, R. (1987) *Visible and Ultra-violet Spectroscopy*, Wiley, ISBN 0 471 91378 2.

George, B. and McIntyre, P. (1987) *Infrared Spectroscopy*, Wiley, ISBN 0 471 91383 9.

Metcalfe, E. (1987) *Atomic Absorption and Emission Spectroscopy*, Wiley, ISBN 0 471 91385 5.

Rendell, D. (1987) *Fluorescence*, Wiley, ISBN 0 471 91381 2.

A recent title on NIR is:

Osborne, B.G., Fearn, T. and Hindle, P.H. (1993) *Practical NIR Spectroscopy: with Applications in Food and Beverage Analysis*, 2nd edn, Longman, ISBN 0 582 09946 3.

# 4 Electrochemical techniques

One of the most widely used analytical methods, and indeed one of the first instrumental methods to be invented, is that of pH measurement. You may well have come across the pH meter with its glass electrode. You may have wondered how glass, which is an insulator and chemically rather inert, could be an electrode and how it could respond to acids and alkalies. Whatever the answer, it shows that minute electrochemical effects can be detected and turned into useful analytical measurements.

Since the advent of the pH meter, a wide variety of other meters, electrodes and methods has been developed, often rivalling other techniques in ease of use and ability to measure concentrations. The use of microprocessors has given a new impetus to making the instruments more adaptable and reliable. One of the latest developments has been to replace the glass electrode with a silicon chip and to make a pH meter the size of a credit card.

On completion of this chapter you should be able to:

- define the term redox reaction;
- give examples of redox reactions;
- recognize relative oxidizing strengths on the basis of electrode potentials;
- describe how electrode systems can be set up and how their potentials are measured;
- describe how electrochemical measurements can be used in qualitative and quantitative analysis.

## 4.1 The reactions which are the basis of electrochemistry

In introductions to chemistry you will come across the fact that although there are many individual chemical reactions there are comparatively few reaction types. Among these types are those in which oxidizing and reducing agents play a part and others where the reagents are acids and bases. These two types of reaction are important and widespread in biological systems.

Typical of the first type is the combustion of carbon to carbon dioxide, and because it requires oxygen we call it an **oxidation**. Many other carbon-containing compounds can be burnt (oxidized) and when they give us useful amounts of energy we call them fuels. Most fuels produce water as well as carbon dioxide when they are oxidized. You can burn glucose in air or oxygen although it is not a convenient fuel as compared with petrol or coal. The process gives out energy in the form of heat and light. Biologically, glucose is used as a fuel and its oxidation to carbon dioxide and water is the same overall conversion as is combustion. It is achieved, however, by a complex series of steps in the metabolism of cells. Nevertheless, the whole process is still oxidation and results in the production of energy, but in contrast to the combustion, much of the energy produced is in the chemical form adenosine triphosphate (ATP). Another process, which is essentially the opposite of oxidation, is known as **reduction**: the name was derived from the process of 'reducing' ores to metals. Such extraction of metals is equivalent to the chemical removal of oxygen from the ore. Biological examples of reduction are the building up of fatty acid chains and the microbiological formation of methane. Although we often concentrate our attention on one or other of the two processes we should note that the oxidation of one substance always requires the reduction of another and vice versa, so all these reactions are often known as **redox reactions**.

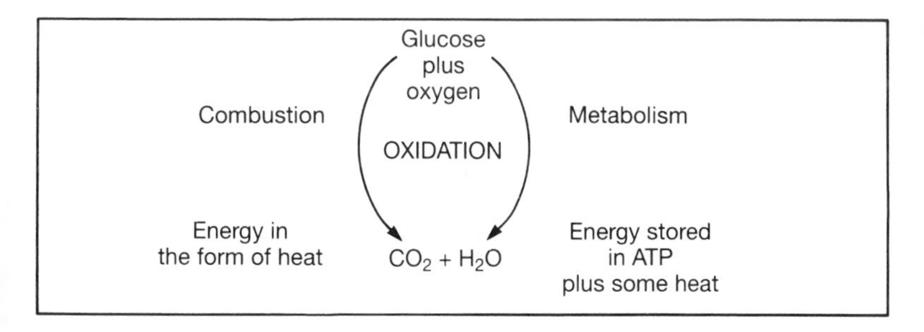


**Figure 4.1** Oxidation.

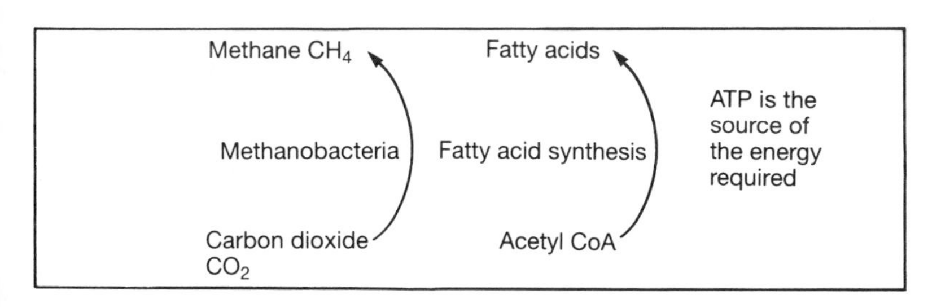


**Figure 4.2** Some biological reductions.

The second type of reaction is between **acids and bases**. When carbon dioxide is carried from the tissues to the lungs, it has an acidifying effect on the blood. This can adversely affect the oxygen-carrying ability of the haemoglobin, and buffering systems come into play to stabilize the pH and minimize the effect (see Box 4.1). Such buffering is also necessary for almost all enzyme reactions, and acids and bases play important roles in other reactions. The body uses acidic conditions to digest proteins and carbohydrates, but alkaline conditions for fats. The bile salts are somewhat alkaline substances (bases) which neutralize the stomach acids before the fats are digested.

**Box 4.1 Acids and bases in biological systems**

What we mean by acids and bases is largely a matter of definition and needs to be related to a solvent. For introductory chemistry and for biology the solvent is water. J.N. Brönsted (a Danish chemist, 1879–1947) and T. Lowry (an English chemist, 1874–1936) both recognized the self-ionization of water:

$$H_2O\ (l) \rightleftharpoons H^+\ (aq) + OH^-\ (aq)$$

and used it to define acids thus:

**acids are substances which produce hydrogen ions in water.**

The process is known as **acid dissociation**, because a covalent molecule dissociates into separate ions. You may also come across the term **proton donor** for an acid since a hydrogen ion in water is essentially a hydrated proton. Also, since one molecule of water may be regarded as more strongly bound to the proton than the rest of the water molecules a hydrogen ion may be called a hydronium ion and written as $H_3O^+$ (aq). The simple chemical representation $H^+$ (aq) is sufficient for our purposes.

The range of concentrations of hydrogen and hydroxide ions that occur in aqueous solutions is extremely wide, and Sørensen devised the pH scale to give a simple numerical reference to these concentrations. He defined pH as the negative logarithm of the hydrogen ion concentration ($pH = -\log [H^+]$).

We also recognize different strengths of acids. **Strong acids are completely ionized in water**, such as hydrochloric acid:

$$\underset{\text{(not present as covalent molecules in water)}}{HCl} \rightarrow H^+\ (aq) + Cl^-\ (aq)$$

Sulphuric and nitric acids are also strong acids in this in this sense. Note that the definition does **not** refer to the **concentration** of the acid solution.

**Weak acids are only partially ionized in water**. There is a continual ionization and recombination of the ions, resulting in a steady state (a dynamic equilibrium). The proportion of ionized molecules depends on the concentration of the acid, but as a rough guide it is generally no more than 1% in a 1 mol $dm^{-3}$ solution. It also depends on the molecular structure of the acid, and the relative strengths of weak acids are important in biological reactions. The **p$K_a$** value measures these relative strengths using the pH scale. It is derived from the **acid dissociation constant** (look for both of these terms in a biological chemistry textbook). Ethanoic (acetic) acid is a typical weak acid and its ionization is represented thus:

$$\underset{\text{about 99\% in this form}}{CH_3COOH\ (aq)} \rightleftharpoons \underset{\text{about 1\% ionized}}{CH_3COO^-\ (aq) + H^+\ (aq)}$$

Most organic acids, and hence most biological acids such as citric acid and and pyruvic acid, are weak acids.

For bases, you might expect the Brönsted–Lowry definition to be a substance that produces hydroxide ions in water, and this is often used when alkalis are the main consideration. Alkalis are soluble strong bases. A more general definition is that

**bases are substances that combine with hydrogen ions.**

Most of the bases that you come across in biology are weak bases (partially ionized) and a better representation of their activity is thus:

$$\underset{\text{about 99\% in this form}}{NH_3\,(aq)} + H^+\,(aq) \rightleftharpoons \underset{\text{about 1\% ionized}}{NH_4^+\,(aq)}$$

Many organic amines such as histamine and heterocyclic substances such as the bases in RNA and DNA behave similarly.

The reaction of an acid with a base is known as **neutralization** since the distinct characters of the acid and base are lost. The product of the reaction is a **salt**. However, a weak acid may be regenerated (liberated) from its salt by the action of a strong acid, which produces a low pH, and a weak base may be liberated from its salt by the action of a strong base, giving a high pH.

**Buffers** are solutions which resist change in pH, and are composed of a weak acid and one of its salts, or a weak base and one of its salts. If you add acid to a buffer, or if hydrogen ions are produced in a reaction, the ionized form of the acid will combine with the hydrogen ions to an extent which keeps the pH nearly constant. Conversely, if you add base, or hydrogen ions are removed by a reaction, the acid will ionize further, making up for the removal. Enzyme reactions are very sensitive to pH and there are many naturally occurring buffers in cells. In biological investigations, therefore, the control of pH is particularly important and buffers are almost always required. Often the change in pH in a buffered reaction is so slight that it is regarded as being constant. The way the pH changes with the concentrations of the acid and salt was determined by L.J. Henderson (an American biochemist and physiologist, 1878–1942), who proposed an equation using the dissociation constant of the acid, and K.A. Hasselbalch (a Danish biochemist), who formulated the logarithmic equation that we use today. It is known as the Henderson–Hasselbalch equation:

$$pH = pK_a + \log\frac{[salt]}{[acid]}$$

A common feature of both types of reaction is that they involve electrically charged particles: ions, electrons and protons. Although this is not always clear when simple chemical equations are written, the fact that such particles are present, and that they can move in the system, gives rise to electrical effects. Such effects may be minute, but can readily be measured with modern microchip circuitry, and enable us to analyse or investigate many aspects of the chemistry of biological samples.

You will need to know the details of such reactions if you are studying biochemistry and you should consult a biochemistry textbook. For the present, we will summarize the important features of redox reactions because they are the basis of electrochemical measurements.

### 4.1.1 Redox reactions

The fundamental mechanism in redox reactions is the **transfer of electrons** from one atom or molecule to another, and the mechanism is a combination of two complementary processes:

- **oxidation,** which is **loss of electrons** by an atom or molecule;
- **reduction,** which is **gain of electrons** by an atom or molecule.

However, the conventional way of writing chemical equations does not always show this clearly. Other ways of recognizing redox reactions use chemical features thus:

| **Oxidation** | **Reduction** |
|---|---|
| gain of oxygen atoms | loss of oxygen atoms |
| loss of hydrogen atoms | gain of hydrogen atoms |
| increase in positive charge | decrease in positive charge |
| decrease in negative charge | increase in negative charge |

Here are some examples of redox reactions, which may be interpreted in terms of these descriptive chemical criteria (mostly shown as reaction schemes rather than balanced equations so that details do not obscure the essentials).

The combustion, or the oxidative metabolism of glucose (see Figure 4.1):

$$C_6H_{12}O_6 + 6O_2 = 6CO_2 + 6H_2O$$

and the souring of wine:

$$\underset{\text{conversion of ethanol}}{2CH_3CH_2OH} \underset{\text{via}}{\xrightarrow{\text{oxygen}}} \underset{\text{ethanal}}{2CH_3CHO} \underset{\text{to}}{\xrightarrow{\text{more oxygen}}} \underset{\text{ethanoic acid}}{2CH_3COOH}$$

An example of reduction is the production of methane by methanobacteria (see Figure 4.2):

$$CO_2 \xrightarrow[\phantom{x}]{\substack{\text{hydrogen} \\ \text{(supplied by NADH)}}} CH_4$$

There are, of course, several steps in this conversion.

In Fehling's or Benedict's test for reducing sugars:

$$\underset{\text{glucose}}{C_6H_{12}O_6} + \underset{\substack{\text{copper(II) ion} \\ \text{in alkaline solution}}}{2Cu^{2+} + 5OH^-} \rightarrow \underset{\substack{\text{gluconic acid} \\ \text{(ionized by the alkali)}}}{C_6H_{11}O_7^-} + \underset{\substack{\text{red ppt of} \\ \text{copper(I) oxide}}}{Cu_2O + 3H_2O}$$

You can see that the glucose gains oxygen and so is oxidized, while the copper changes from a 2+ ion to a 1+ ion in the copper(I) oxide and is reduced. In order to show the reactions more clearly in terms of electron transfer we use the concept of 'half-reactions'.

### 4.1.2 Half-reactions

Although an oxidation cannot take place without a reduction occurring, it is often convenient to consider the processes of oxidation and reduction separately. In order to write them down and compare them, we use a convention which puts the oxidized form of a substance on the left of an equation together with the electrons and any other substances that are required for the process; the reduced form appears on the right. The equation only represents half of the overall redox process and thus is known as a **half-equation** and the process is a **half-reaction**. Also the half-equation is shown as being reversible since the half-reaction can be made to proceed either way by applying suitable conditions. Thus reduction is the process going left to right in the half-equation and oxidation is right to left:

$$\text{oxidized form of the substance} + \text{electrons} \rightleftharpoons \text{reduced form of the substance}$$

which can be put in an abbrieviated form thus:

$$\text{ox} + n\text{e}^- \rightleftharpoons \text{red}$$

where $n$ is the number of electrons.

You will see in Table 4.1 that the examples given there are written in this half-reaction form. In any complete redox reaction, as one substance becomes oxidized another becomes reduced, so we will require two of these half-equations to show one reaction. Any two half-equations can be combined to give a reaction which is at least theoretically possible.

Substances which cause oxidations are known as **oxidizing agents** or **oxidants** and any of the chemical species on the left of a half-equation can be an oxidant. Any on the right can be a **reducing agent** or **reductant**. As we mentioned above, all redox processes are reversible and so what actually happens in a reaction is generally decided by which of the two half-reactions is the stronger oxidizing system (see Section 4.3). However, factors such as concentration (including pH) and solubility may also influence the final outcome of the reaction.

To make a redox reaction occur you must combine the components of two half-reactions in one system. You can do this just by mixing the components in a beaker, or you can keep the two half-reactions physically separate but connect them by means of electrodes and conductors. For instance, if we consider the reaction between copper(II) ions and metallic zinc the overall reaction is that the copper(II) ions oxidize the zinc atoms to zinc ions and become reduced to copper atoms in the process:

Figure 4.3 illustrates this.

| $Cu^{2+}$ | + | Zn | → | Cu | + | $Zn^{2+}$ |
|---|---|---|---|---|---|---|
| ions in solution | | metal atoms | | metal atoms | | ions in solution |
| blue | | | | brown deposit | | colourless |

When we look at this in terms of half-reactions, we say that the more strongly oxidizing half-reaction:

$$Cu^{2+} + 2e^- \rightarrow Cu$$

**Table 4.1** Some common and some biologically important half-reactions

| *Chemical systems* | *Standard electrode potentials (pH = 0) $E^0$ (V)* | *Comments* |
|---|---|---|
| | Most strongly oxidizing | |
| $F_2 + 2e^- \rightleftharpoons 2F^-$ | +2.85 | The strongest oxidizing agent |
| $H_2O_2 + 2H^+ + 2e^- \rightleftharpoons 2H_2O$ | +1.77 | |
| $MnO_4^- + 8H^+ + 5e^- \rightleftharpoons Mn^{3+} + 4H_2O$ | +1.52 | |
| $Cl_2 + 2e^- \rightleftharpoons 2Cl^-$ | +1.36 | |
| $O_2 + 4H^+ + 4e^- \rightleftharpoons 2H_2O$ | 1.22 | |
| $NO_3^- + 2H^+ + 2e^- \rightleftharpoons NO_2 + H_2O$ | +0.81 | |
| $Fe^{3+} + e^- \rightleftharpoons Fe^{2+}$ | +0.77 | Compare with the $E_0'$ value below |
| $O_2 + 2H^+ + 2e^- \rightleftharpoons H_2O_2$ | +0.68 | |
| $Cu^{2+} + 2e^- \rightleftharpoons Cu$ | +0.34 | As used in Fehling's and Benedict's tests |
| $Cu^{2+} + e^- \rightleftharpoons Cu^+$ | +0.16 | The defined zero of the $E^0$ scale |
| $2H^+ + 2e^- \rightleftharpoons H_2$ | 0.00 | Reactive metals are reducing agents |
| $Zn^{2+} + 2e^- \rightleftharpoons Zn$ | −0.76 | Reactive metals are reducing agents |
| $Na^+ + e^- \rightleftharpoons Na$ | −2.71 | |
| | Most strongly reducing | |
| *Biological systems* | *Biological standard potentials (pH = 7) $E_0'$ (V)* | *Comments* |
| | more oxidizing | |
| $O_2 + 2H^+ + 2e^- \rightleftharpoons H_2O_2$ | +0.30 | Used by several oxidase enzymes |
| Cytochrome c $Fe^{3+} \rightleftharpoons Fe^{2+}$ | +0.25 | |
| Oxidized DPIP $\rightleftharpoons$ DPIP | +0.25 | Not a biological system as such but used in the titration of vitamin C |
| Oxidized vit. C $\rightleftharpoons$ vit. C | +0.06 | |
| Pyruvate $\rightleftharpoons$ lactate | −0.19 | Catalysed by lactate dehydrogenase |
| Ethanal $\rightleftharpoons$ ethanol | −0.02 | Catalysed by alcohol dehydrogenase |
| FAD $\rightleftharpoons$ $FADH_2$ | −0.22 | May act as oxidizing system if concentration of FAD is high |
| $NAD^+ \rightleftharpoons$ NADH | −0.32 | May act as oxidizing system if concentration of $NAD^+$ is high |
| | more reducing | |

Much fuller lists can be found in biochemistry and analytical textbooks.

causes the less strongly oxidizing half-reaction

$$Zn \rightarrow Zn^{2+} + 2e^-$$

to become oxidized (this latter half-reaction proceeds in reverse as compared with its presentation in Table 4.1, i.e. right to left in the conventional half-equation). Thus we can say that the $Cu^{2+}$ is the oxidizing agent and, from the complementary point of view, the Zn is the reducing agent.

Copper(II) ions are also the oxidizing agent in Fehling's and Benedict's tests, which we mentioned earlier, although in these reactions they are reduced to copper(I) ions which are precipitated in the form of the red copper(I) oxide. The half-equations can be written:

$$2Cu^{2+} + 2OH^- + 2e^- \rightarrow Cu_2O + H_2O$$

(which is, in essence, $Cu^{2+} + e^- \rightarrow Cu^+$)

Can you see that these two half-equations combine to give the full equation given in Section 4.1.1.

and

$$C_6H_{12}O_6 + 3OH^- \rightarrow C_6H_{11}O_7^- + 2H_2O + 2e^-$$

The half-reactions for very strong oxidizing agents always proceed left to right in the conventional form of the equation if present in most biological systems. Conversely, very strong reducing agents go right to left. Those which are of intermediate strength may go either way depending on the strength of the other half-reaction. Let us consider a solution containing iron(II) ions in contact with air. We find from common experience that oxygen from the air oxidizes acidified iron(II) solution to give iron(III), which is why we cannot store iron(II) solutions for more than about a day when they are needed for precise analytical work. We can find the relevant half-equations in Table 4.1:

$$O_2 + 4H^+ + 4e^- \rightleftharpoons 2H_2O$$

$$Fe^{3+} + e^- \rightleftharpoons Fe^{2+}$$

The iron half-reaction must be the less strongly oxidizing since, as we have just said, it is the iron(II) ions which become oxidized and so it has to be written 'back to front' when it is combined with the oxygen half-equation. Also, the overall reaction cannot have any spare electrons, so we must combine the first half-equation with four times the second, which gives:

$$O_2 + 4H^+ + 4Fe^{2+} \rightarrow 4Fe^{3+} + 2H_2O$$

In this reaction, the iron(III) ion is not a strong enough oxidizing agent to cause any oxidation in the system. However, if the iron half-reaction were to be combined with a tin(IV)/tin(II) half-reaction, the iron(III) ion would be the stronger oxidizing agent and tin(II) would be oxidized to tin(IV):

$$2Fe^{3+} + 2e^- \rightarrow 2Fe^{2+}$$

(more strongly oxidizing than $Sn^{4+}$)

$$Sn^{2+} \rightarrow Sn^{4+} + 2e^-$$

(less strongly oxidizing than $Fe^{3+}$, or more reducing than $Fe^{2+}$),

so the overall reaction is:

$$2Fe^{3+} + Sn^{2+} \rightarrow 2Fe^{2+} + Sn^{4+}$$

We can say that iron(III) oxidizes tin(II) to tin(IV) or that tin(II) reduces iron(III).

Thus we should note that whether a substance acts as an oxidizing agent or as a reducing agent depends on what other substances it is reacting with and their concentrations (see Section 4.3). In biological systems this occurs with enzymes using the cofactors $NAD^+/NADH$ and $FAD/FADH_2$.

## 4.2 Electrode systems

So far we have shown that electrons are transferred in redox reactions, but we have given little indication as to how they form the basis of electrochemistry. The chemical power that drives the reactions is turned into electrical power when we set up an electrode system such as is shown in Figure 4.3. We have then created an **electrochemical cell** which will produce voltages and currents, depending on the conditions in the electrode compartments, both of which give useful information in analysis. In biology the study of nerve conduction shows that small voltages are produced across nerve membranes, and the electric eel is an example of how biological systems can produce a current.

In electrochemical circuits there are parts where electrons move (the wires and the meter) and parts where ions move (the solutions and the salt-bridge). These parts do not actively participate in the reactions that produce the electrochemical effects; they just provide electrical conduction. The electrodes, however, are the interfaces where electron exchange takes place. At the surface of the electrode ions may be formed or discharged, or undergo a change in charge. Look again at Figure 4.3. You should note that each electrode corresponds to one half-reaction and there must be two electrodes in order to establish a complete circuit. At the negative electrode (anode) electrons are lost from ions, molecules or atoms (oxidation). This provides a supply of electrons to the positive electrode (cathode) where electrons are supplied to atoms, molecules or ions and the process is reduction.

The terms anode and cathode can be confusing since, in electrolysis, they carry the opposite charges from those stated here. However, whether we are discussing electrolytic or electrochemical cells, oxidation always occurs at the anode and reduction at the cathode.

### 4.2.1 General-purpose electrodes

*Metal electrodes*

These simply consist of a metal surface in contact with a solution. The atoms of the electrode metal are involved in the half-reaction, and the solution which is in contact with that electrode must contain ions of the metal. The electrodes $Zn/Zn^{2+}$ and $Cu/Cu^{2+}$ are shown in Figure 4.3. Very reactive metals, such as sodium, will chemically attack the water and so it is not possible to construct a simple electrode from them.

The '/' symbol denotes the electrode surface.

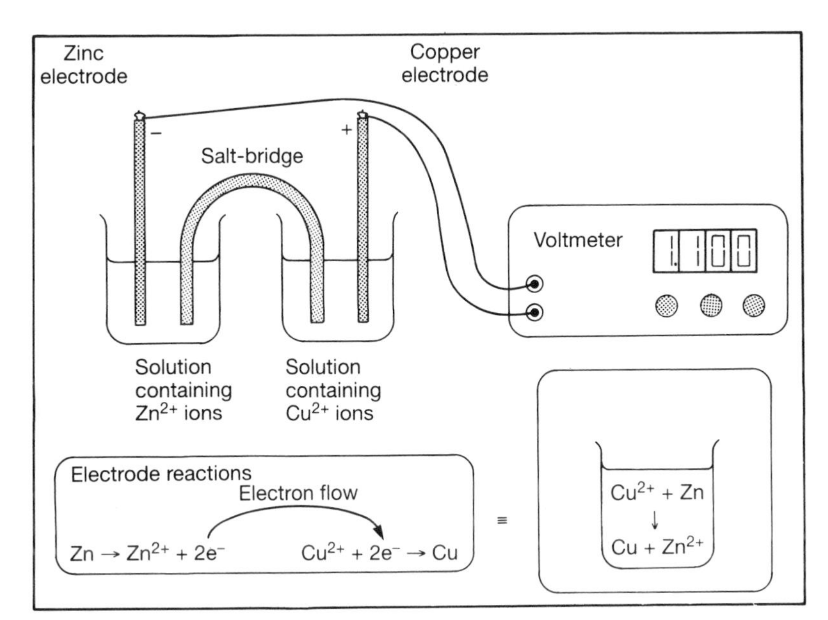


**Figure 4.3** Typical simple electrode set-up.

*Inert electrodes*

In contrast, metals of very low reactivity can be used as inert electrodes, allowing other electrochemical processes to occur without themselves being involved chemically.

**Platinum electrodes** are widely used in this way, particularly for the more complex half-reactions. For instance, they allow the iron(II)/iron(III) system to produce or use up electrons, and they can be used in half-reactions involving gases (see Figure 4.4).

**Carbon electrodes**: carbon is not a metal and it does not form simple ions, but the form known as graphite conducts electricity and can be used as a material for inert electrodes. Its performance as an electrode is very dependent on the chemical state of its surface. A particular modification, 'glassy carbon', is used in several applications.

*Gas electrodes*

In order to bring gases into electrical contact with the systems we are considering, they have to be adsorbed onto a suitable surface. Platinum is suitable when its surface has been made catalytically active by the deposition of a layer of 'platinum black'. A sleeve channels the gas down to the electrode, as shown in Figure 4.4, or a bubbler can release the gas just below the level of the electrode. The electrode can be used for gases such as $H_2$, $O_2$ and $Cl_2$. The solution must contain ions derived from these gases, for example $H^+$, $OH^-$, $Cl^-$.

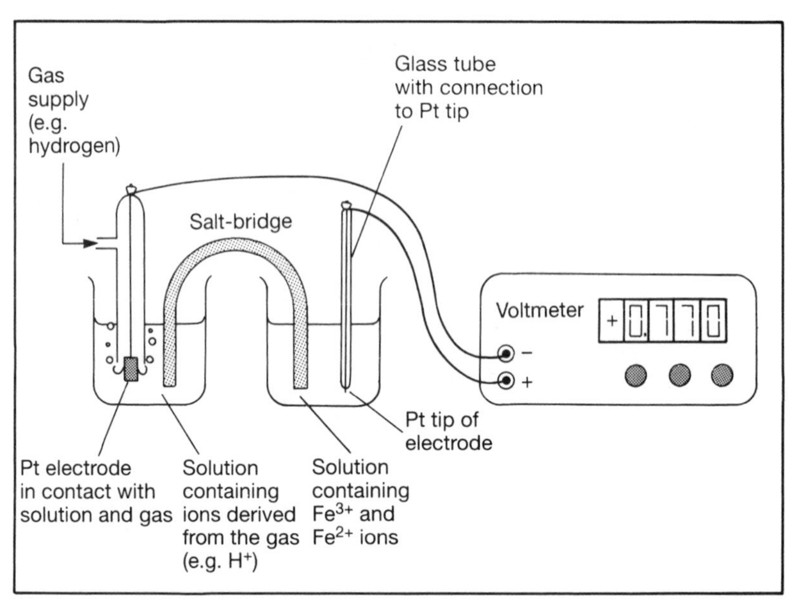


**Figure 4.4** General redox and gas electrodes.

## 4.2.2 Reference electrodes

You often have to measure the voltage developed by an electrode and it is done in conjunction with a **reference electrode** which gives its own precisely known voltage that does not depend on the conditions in the test solution. The ultimate reference is the hydrogen electrode – one of the gas electrodes mentioned in the previous paragraph – but there are several difficulties in handling hydrogen (its flammability, the need to have constant pressure and so on) which make it inconvenient for routine use. Secondary reference electrodes have been devised which are easy to handle. Their interior contains a saturated solution making electrical contact with the test solution through a narrow porous plug which acts as the salt-bridge (see Figure 4.5). Diffusion of ions through the plug is virtually zero, and so the internal electrolyte composition is not significantly affected. Thus these electrodes give voltages which depend only on temperature.

The two most commonly used reference electrodes are silver/silver chloride and 'calomel', which is mercury(II) chloride.

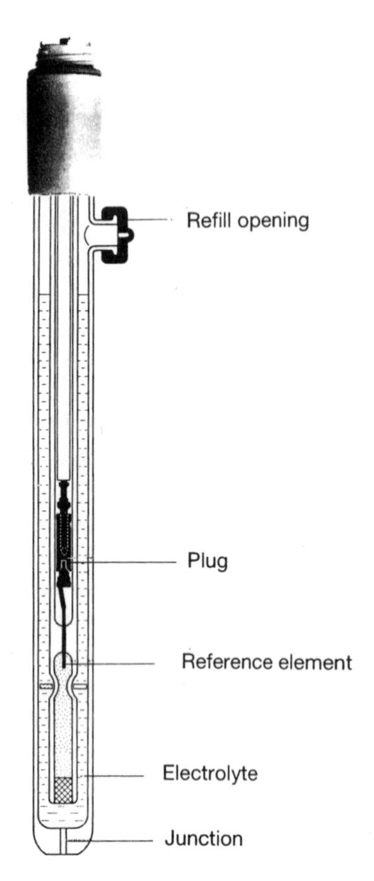


**Figure 4.5** Silver/silver chloride reference electrode. (Courtesy Mettler-Toledo Ltd.)

## 4.2.3 Specialized electrodes

*Ion-selective electrodes*

In contrast to reference electrodes, **ion-selective electrodes** (ISEs) do respond to changes in the test solution, but ideally only to one particular ion. In practice, some of these electrodes are affected by presence of other ions but usually to a much lesser extent.

*pH electrodes.* The most widely used ISE is the glass pH electrode (Figure 4.6). Its membrane is a sensitive area in the glass (usually a bulb at the end of the electrode) which responds only to the concentration of hydrogen ions in the solution in which it is immersed. It gives a voltage which is converted to a pH reading by the electronics of the meter.

Glass pH electrodes have been in use for several decades. A wide variety of shapes and styles has been developed to suit all sorts of applications (Figure 4.7). The quality of meat depends on the pH and so it is tested with a sharp narrow electrode that can penetrate the samples. Alternatively, you may need to know the pH of a surface such as a gel for growing microorganisms. An electrode for this has a flat area rather than a bulb or a sharp point. Although very effective, glass pH electrodes suffer from fragility, and they can be unreliable if they dry out. pH electrodes have been devised using other materials, such as glassy carbon mentioned under 'inert electrodes' and ion-exchange resins, while the latest development is to incorporate a pH-sensitive area into a micro-chip, which produces a rugged electrode that does not have to be kept moist.

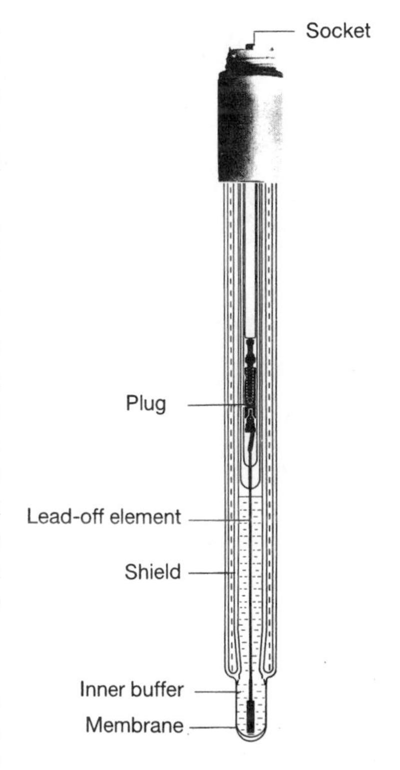


**Figure 4.6** A typical glass pH electrode. (Courtesy Mettler-Toledo Ltd.)

*Other types of solid membrane ion-selective electrode.* In common with the pH electrode, other ISEs have one portion of the electrode surface sensitive to the ion selected for and this is called the 'membrane'. There is a variety of materials from which the membrane may be composed and some of the more common ones are indicated in Table 4.2.

**Glass** can be modified to be sensitive to ions other than hydrogen, particularly those of the alkali metals lithium and sodium. Some **crystalline** ionic materials can be used as long as they contain the ion to be selected for (chloride or fluoride, for example) and are not soluble in the solutions to be tested.

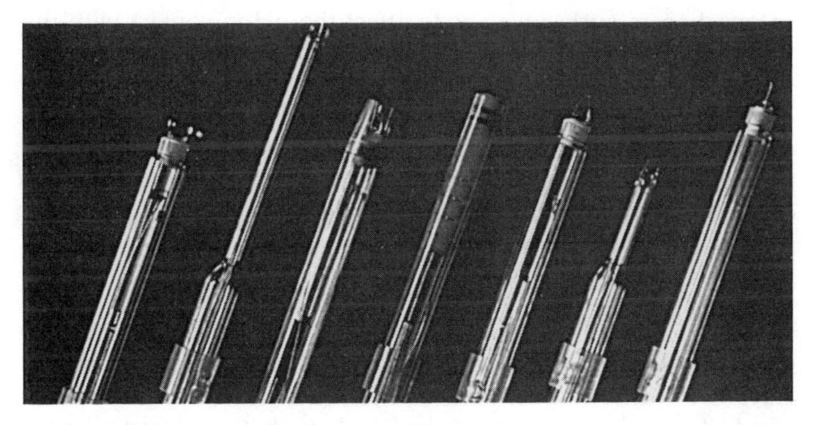

**Figure 4.7** A variety of pH electrodes. (Courtesy Jenway, UK.)

**Table 4.2** Types of ion-selective electrode

| *Membrane material* | *Selective for* | *Interfering ions* |
|---|---|---|
| Solid | | |
| Glass (various sorts) | $H^+$ | |
| | $Li^+$ | |
| | $Na^+$ | $H^+$, $Li^+$ |
| Crystalline | | |
| $LaF_3$ | $F^-$ | $OH^-$ |
| $Ag_2S$/chloride and other halides similarly | $Cl^-$ | $Br^-$, $I^-$, $CN^-$ |
| Liquid | | |
| Plastic film impregnated with a selective liquid ion-exchange material | | |
| | $K^+$ | $Rb^+$, $Cs^+$, $NH_4^+$ |
| | $Ca^{2+}$ | $Zn^{2+}$, $Pb^{2+}$ |
| | and several other divalent metal ions | |
| | $NO_3^-$ | $ClO_3^-$, $ClO_4^-$ |
| | and many other anions | |

*Liquid membrane ion-selective electrodes.* Alternatively, ion-exchange materials (ionophores) which form a complex with particular ions but not others are used. These materials may be liquids, in which case they are absorbed into a thin layer of porous plastic.

Many common ions can be analysed using ISEs and a summary of the range and types is given in Table 4.2.

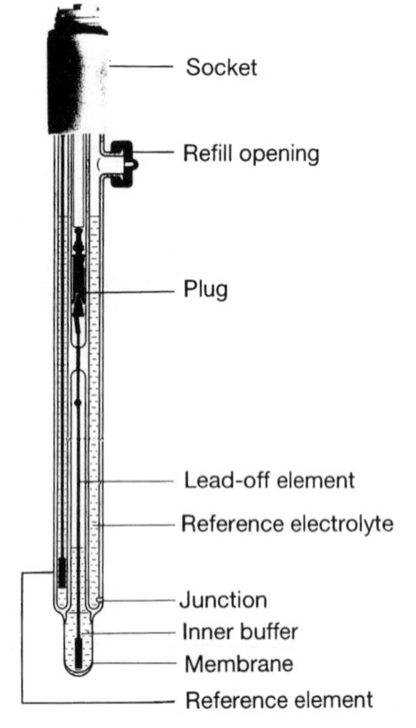


**Figure 4.8** A combination pH electrode. (Courtesy Mettler-Toledo Ltd.)

### *Adapted ion-selective electrodes*

*Gas-sensing electrodes.* These are not the same as the gas electrodes described above where a supply of the gas is required. Here the membrane of a pH electrode is covered with another membrane which allows the passage of gases but not other substances. The electrolyte in the system must respond to the gas in question in such a way as to affect the pH in contact with the glass membrane. This usually means that the gases detected are acidic ($CO_2$, $SO_2$) or alkaline ($NH_3$, some simple amines). If the solution contains an enzyme reaction which produces or uses up gas the adapted electrode can be used to monitor the activity of enzyme or it can be used to measure the concentration of the substrate since the enzyme activity depends on this. Some examples are given in Table 4.3.

### *Combination electrodes*

As we noted earlier, all electrochemical measurements have to have two electrodes. For pairs of electrodes which are very widely used, it may be more convenient to have both electrodes incor-

**Table 4.3** Some adapted ion-selective electrodes

| *Substrate* | *Type of electrode* | *Biocatalyst* |
|---|---|---|
| Ethanal (acetaldehyde) | pH (glass) | Aldehyde dehydrogenase |
| Glutamate | Gas-sensing ($CO_2$) | Bacteria (containing the enzyme glutamate decarboxylase) |
| Serine | Gas-sensing ($NH_3$) | Serine deaminase |

From *International Laboratory*, Vol. 13, no. 6, July/August 1983, 24–32. Courtesy of Mark A. Arnold.

porated into the same assembly. This gives a **combination electrode** which looks like a single electrode (although it is actually two) and is easier to handle than two separate electrodes. This type of combination is done with the general-purpose platinum redox electrode and the pH electrode (Figure 4.8). In fact, any pair of electrodes could be designed as a combination electrode, but most are much less widely used than the platinum or the pH electrodes, so the benefits of bulk manufacture are not obtained and they are not economic to produce.

## 4.3 Electrode potentials and their measurement

We will start this section by discussing the term 'potential'. Although we may talk about '**a** potential', we can only actually measure **differences** between two potentials and any voltage is actually the value of a potential difference. In ordinary voltage measurements we usually define the earth or ground as being zero volts, so the scale of voltage is numerically equal to the potential difference between a point in a circuit and the ground.

When we measure voltages generated by electrodes we find that they depend on how much current is flowing in the circuit. If we allow a larger current to flow the voltage drops, but the smaller the current the greater the voltage, up to a limiting value dependent on the electrode system. If we take this to its logical extreme, we will measure the maximum voltage when the current in the circuit is zero (in practice we would say 'vanishingly small'). This voltage is the **electrode potential** and it is not affected by trivial factors which come into play when currents are flowing, such as the size of the electrodes. For electrode measurements to be comparable from one experiment to another we need to refer to these potentials.

Measuring potentials allows our attention to be focused on the fundamental aspects of the system. The methods used to measure potentials with zero current flowing include potential matching and the use of meters with very high-impedence circuitry whose details can be found in physics and electronics textbooks.

It is important that you do understand the difference between voltage and current. You may need to consult a physics textbook for this but an approximation is that voltage is a measure of the electrical 'pressure', while current (amps) is a measure of the electrical flow. It is wrong to say '. . . 100 volts running through a circuit . . .' or '. . . we applied a current to . . .'. You may apply a voltage to a circuit and a current will run through it. You will frequently come across such phrases in the media!

Remember that the potentials are developed by the electrodes, they are not applied voltages as in electrolysis. The first record of the generation of electricity by a chemical system was made by Luigi Galvani (an Italian biologist, 1737–1789). His famous observation was of the twitching of a frog's leg which was caused by the leg being in contact with both an iron frame and some dissection implements made from a different metal. These implements and the frame acted as electrodes and the leg completed the circuit. You may come across the occasional use of the term 'Galvanic cells' for the systems we are describing.

### 4.3.1 Measurements involving electrode potentials (potentiometry)

In order to set up an electrochemical circuit you need two electrodes and suitable connections between them, as shown in Figures 4.3 and 4.4. In part of the circuit where electrons are flowing ordinary wires are suitable, but where ions are moving (in the solutions surrounding the electrodes) the connection has to be via a 'salt-bridge', which is a tube containing a gel of agar and potassium chloride. This salt is chosen because its ions are similar in size and move with equal but opposite velocities, which avoids the production of other voltages due to unequal movement. The voltmeter must be of the potentiometer or high-impedance type so that virtually no current flows.

### 4.3.2 The electromotive force of an electrochemical cell

You can connect any two electrodes in the manner shown in Figures 4.3 and 4.4 to produce an electrochemical cell. It is effectively a battery, but whether it would be capable of powering a torch or a transistor radio is another matter. In potentiometry the capability of a cell to produce a current, and hence some power, is not under consideration. Under zero current conditions the voltage that we measure is the potential difference between the two electrodes and is often known as the **electromotive force (emf)** of the cell.

At the turn of the last century such cells were used to provide the power for electric door-bells. However, these cells were not readily portable because of the liquid contents. A distinct advance was the development of the so-called dry cell, which is the everyday torch battery. Today there is a variety of similar cells and the race is on to produce a satisfactory fuel cell that could power a car.

For a cell composed of zinc and copper electrodes, we can summarize the system shown in Figure 4.3 as:

$$Zn^{2+} / Zn // Cu / Cu^{2+}$$

in which / indicates the electrode surface and // indicates the salt-bridge. It is the convention to put the anode (the electrode where oxidation is taking place) on the left. We find that, under ideal conditions:

$$\text{emf of this cell} = 1.1\,\text{V}$$

### 4.3.3 Establishing the scale of electrode potentials

We generally use the symbol $E$ to denote an electrode potential. Thus we can write:

$$\text{emf} = E_c - E_a$$

Remember that we measure potential differences, hence the subtraction of the two $E$ values.

where $E_c$ represents the potential of the cathode (which will be the positive electrode and is connected to the terminal labelled '+' on the voltmeter) and $E_a$ the anode (the negative electrode).

Since we always have to have two electrodes in order to make a measurement, we cannot measure the potential of a single electrode itself and there is no absolute scale of electrode potentials. We can, however, retain the concept that there is a potential developed by each electrode and we use a scale which has a **defined** zero instead of an absolute one. So, when we refer to the potential of a particular electrode we should remember that it always means the potential with respect to the defined zero.

The lack of an absolute zero applies to all sorts of potentials, not just electrical ones. Potential energy is always measured from some arbitrary zero. An analogy is found in geography; heights are measured from the arbitrary zero of sea-level – there is no absolute zero of height. But can you see why the measurement of temperature does not provide a complete analogy?

### 4.3.4 The reference zero of electrode potentials; the hydrogen electrode

The half-reaction occurring at a hydrogen electrode has been chosen as the arbitrary zero of electrode potentials. Hydrogen is the simplest atom and is, in a sense, the meeting point of redox and acid–base reactions so it is a particularly important element. The half-equation may be written as:

$$2H^+ + 2e^- \rightleftharpoons H_2$$

We need to use particular conditions of temperature, pressure and concentration for this zero to be precise. These conditions are known as the **standard state** and are detailed below (Section 4.3.6).

### 4.3.5 Determining the values of electrode potentials

*Using a hydrogen electrode*

This has been done in some cases but it is technically quite difficult. Great care is needed for both precision and for safety, as was mentioned in the description of the electrode (Section 4.2.1). In these measurements the hydrogen electrode is always connected to the negative terminal of the voltmeter, whether it is the less positive electrode or not. Then:

$$\text{emf} = E_{\text{test}} - E_{\text{H}^+/\text{H}_2}$$

which, under standard state conditions (see below) becomes:

$$\text{emf} = E_{\text{test}} - 0.0 = E_{\text{test}}$$

so the emf measured is the electrode potential of the test electrode.

*Using some other reference electrode*

The difficulty of routinely using a hydrogen electrode is overcome by using one of the reference electrodes described earlier. In this situation we can use the set-up as shown in Figure 4.9. The emf of the cell is related to the individual electrode potentials thus:

$$\text{emf} = E_{\text{test}} - E_{\text{ref}}$$

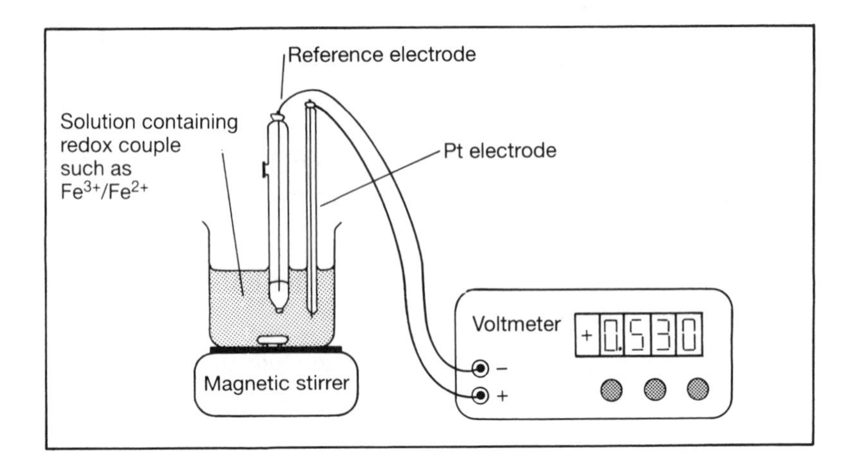


**Figure 4.9** Using a reference electrode.

We can rewrite this to give an equation for the potential of the electrode we are using:

$$E_{test} = \text{emf} + E_{ref}$$

So if we measure an emf and find it to be, say, 0.53 V and we know that the reference electrode develops 0.24 V at the temperature of the experiment, we can calculate that the electrode potential of the test electrode is:

$$E_{test} = \text{emf} + E_{ref} = 0.53 + 0.24 = 0.77\,\text{V}$$

### 4.3.6 Standard electrode reduction potentials

Various factors affect the magnitude of an electrode potential. In order to make comparisons between work in different experiments and from different laboratories we need a set of defined conditions by which we can standardize the values we determine. The **standard state** conditions are widely used for this purpose and when applied to electrode potentials give **standard electrode reduction potentials** (symbol $E^{\ominus}$); the inclusion of the word 'reduction' indicates that it is reduction occurring at the test electrode when a positive emf is obtained. The defined conditions are:

- concentration (or more precisely 'activity'; see Box 4.2) for each ionic species involved in the reaction = $1\,\text{mol}\,\text{dm}^{-3}$;
- pressure of any gases involved = 1 atm (101.3 kPa);
- temperature = 298 K (25°C) unless otherwise specified.

**Box 4.2 Concentrations and the concept of 'activity'**

At low concentrations the dissolved particles in solutions are relatively far apart and so behave virtually independently, which gives rise to ideal behaviour. However, at higher concentrations the particles spend a much greater proportion of their time in close proximity. In this case the interactions between the particles become significant and deviations from ideal behaviour occur.

To allow for this when dealing with concentrations, the concept of **activity** has been devised. It may be described as the 'effective concentration' of the solution. This is the concentration which would be present if the solution were behaving in an ideal manner. In most cases the activity is significantly less than the actual concentration except for very dilute solutions.

Activity is related to the actual concentration by an **activity coefficient** ($\gamma$) thus:

| activity | = activity coefficient × | molar concentration |
|---|---|---|
| ($mol\ dm^{-3}$) | (a number) | ($mol\ dm^{-3}$) |

or

$$a = \gamma \times c$$

The value of the activity coefficient ($\gamma$) varies with concentration and is usually less than 1 for concentrations of more than about 0.1 $mol\ dm^{-3}$, but is approximately equal to 1 for concentrations lower than this. So, for solutions whose concentrations are less than about 0.1 $mol\ dm^{-3}$, which is often the case in biological studies, we can say that values of concentrations and activity are approximately equal.

Since the standard state conditions are fixed, the $E^{\ominus}$ values are fixed and you can find lists of them in biochemistry and physical chemistry textbooks. We listed a few of the biologically more important ones in Table 4.1.

In general you will find it useful to remember that:

- $E^{\ominus}_{H^+/H_2}$ = **0.0 V**, by definition;
- **positive values of** $E^{\ominus}$ mean **more strongly oxidizing than** $H^+/H_2$ (or less strongly reducing);
- **negative values of** $E^{\ominus}$ mean **less strongly oxidizing than** $H^+/H_2$ (or more strongly reducing).

Of course, you will not always want to make a comparison with the hydrogen half-reaction, so **for any pair of half-reactions**:

- **more positive** (or less negative) $E^{\ominus}$ = **more** strongly **oxidizing** (less strongly reducing);
- less positive (or **more negative**) $E^{\ominus}$ = less oxidizing (**more** strongly **reducing**).

## 4.3.7 Non-standard scales for electrode potentials

In the previous section we stated that the definition of standard state included the activity value being one unit. However, the determination of activity values is not straightforward and so there is an alternative scale of values for electrode potentials based on concentrations being 1 $mol\ dm^{-3}$ whether the activities are or not. This scale is given the symbol $E_0$. It is more convenient for routine work which does not require fundamental investigation into electrochemical phenomena.

Even when the $E_0$ scale is used, however, the implication is that the values obtained refer to conditions in which the pH of the solution is 0 ($[H^+] = 1\,mol\,dm^{-3}$), which is far lower than anything obtained in biological systems. So another alternative is used in biological work which is more in line with many biological situations; most cellular pH values are close to 7. The $\boldsymbol{E_0'}$ scale has concentration = $1\,mol\,dm^{-3}$ for all species except for hydrogen ions, where concentration = $10^{-7}\,mol\,dm^{-3}$ (which gives pH 7). A list of some typical $E_0'$ values is given in Table 4.1.

Some typical pH values that are found in nature are lemon juice, pH 2.3; saliva pH 6.5; bile pH 8; but blood and cell pH is always close to 7.2.

Whichever scale is used, the relative oxidizing (or reducing) strengths are given by the magnitude of the $E$ values as described above. Thus we can see from Table 4.1 that oxygen ($E_0' = +0.3\,V$) will oxidize vitamin C ($E_0' = +0.06\,V$) when the $E_0'$ conditions are met. However, $E_0'$ values can be misleading when applied to cellular systems, since the concentrations in the immediate vicinity of an enzyme may differ from those in bulk solution. Such variations allow the cell to control enzyme activity. Also we can note that the redox cofactors $NAD^+$/NADH and FAD/$FADH_2$ have $E_0'$ values that are neither strongly oxidizing nor strongly reducing. Slight changes in conditions can change them from oxidizing agents to reducing agents, which enables a cell to respond readily to changes in its surroundings.

### 4.3.8 Factors affecting electrode potentials

What we have just said might imply that we have to make all our measurements at precisely defined conditions and cannot use the scales for other conditions. This is not the case and we can use variation of conditions to form the basis of electrochemical analysis. First, let us list the factors that can affect the value of an electrode potential:

- the identity of the substances in the half-reaction;
- the number of electrons in the half-equation;
- the nature of the electrode material;
- the concentrations of the substances in the solution;
- the pressure of any gases involved in the half-reaction;
- the temperature.

This may seem like a rather forbidding list and the overall effect of all the factors is summed up by the **Nernst equation** (see Box 4.3). However, we can simplify matters to a large extent by recognizing that the first two factors are fixed as soon as we specify the half-reaction of interest. Also, it is unlikely that we will wish to change the electrode material from one experiment to another, so that can also be regarded as constant once it is specified. Again, in many biological systems the temperature is constant (or approximately so), which removes one more variable. Finally, when we recognize that pressure affects the concentrations of gases in solution, we find that we are chiefly concerned with the effects of concentration on the electrode potential.

**Box 4.3 The Nernst equation**

$$E = E^{\ominus} + \frac{RT}{nF}\ln\frac{[\mathrm{ox}]}{[\mathrm{red}]}$$

This always seems like a very complex equation when you first see it, but by examining each term we can show that it relates directly to the factors we have just been discussing:

$E$ = the electrode potential developed in the conditions under consideration;
$E^{\ominus}$ = the standard electrode potential;
$R$ = the 'universal gas constant'. This is involved in many relations between a system and its energy (potential). It always has the value 8.314 J $K^{-1}$ $mol^{-1}$;
$T$ = the temperature, which must be measured on the absolute scale (add 273 to the centigrade temperature);
$F$ = the Faraday constant. It is the electrical charge on 1 mol of electrons and has the value 96 500 C $mol^{-1}$.

The remaining terms refer to the half-equation ($\mathrm{ox} + ne^- \rightleftharpoons \mathrm{red}$) so:

$n$ = the number of electrons in the half-equation;
[ox] = the concentration (or activity, see Box 4.2) of the more oxidized (less reduced) form of the substance;
[red] = the concentration (or activity) of the less oxidized (more reduced) form.

Until the advent of the scientific calculator the use of natural logarithms (ln or $\log_e$) was not routine. The usual procedure was to substitute $2.303 \times \log_{10}$ for ln. This gives the same value as ln; try it on your calculator and see! You may still come across the $\log_{10}$ version of the Nernst equation in some texts:

$$E = E^{\ominus} + \frac{2.303\,RT}{nF}\log\frac{[\mathrm{ox}]}{[\mathrm{red}]}$$

Yet again, if we substitute the numerical values for $R$ and $F$ and assume a temperature of 298 K the equation may be presented in the form:

$$E = E^{\ominus} + \frac{0.059}{n}\log\frac{[\mathrm{ox}]}{[\mathrm{red}]}$$

You should note that all these forms are different versions of the **same** equation.

When you plot a calibration graph for an ISE you should get a gradient of $RT/nF$ if you use natural logs. If we assume you are working at 25°C, you can calculate from the values given above that the gradient of the graph should be +0.0257 V (25.7 mV) for an ion with a single positive charge and −0.0257 V (−25.7 mV) for a singly charged negative ion. If you used base 10 logs the gradients are ±0.0591 V (±59.1 mV).

Michael Faraday was the great English chemist and physicist who discovered electromagnetic induction and used electrolysis to isolate some reactive elements such as sodium.

## 4.3.9 Changes in concentrations

Looking at the Nernst equation you can see that it is the ratio of the concentrations of the oxidized form, [ox], and the reduced form, [red], that affects the value of $E$. Theoretically the ratio can range from zero (when [ox] = 0) to infinity (if [red] = 0). This leads to the logarithmic term ranging from infinitely negative (ln 0, which would imply from an infinitely strong reducing agent) to infinitely positive (an infinitely strong oxidizing agent)! In practice, there are always significant concentrations of both forms so the range is much more restricted, but you can still make the generalization that a greater proportion of the oxidized form indicates a more oxidizing system and smaller proportion indicates a more reducing system.

We can apply this principle to a well-known effect which has a metabolic explanation in terms of the relative concentrations of oxidized and reduced forms. The cofactor nicotinamide adenine dinucleotide may either act as an oxidizing agent, in which case its oxidized form, $NAD^+$, removes hydrogen atoms from the substrate, or it may be a reducing agent when the reduced form, NADH, supplies hydrogen atoms to the substrate. For instance, the normal metabolism of pyruvate is for it to be converted to acetyl coenzyme A, which is taken into the citric acid cycle. In turn, the cycle produces NADH, which is converted to $NAD^+$ by the respiratory chain. In over-exercised muscle, however, the rate of production of $NAD^+$ is not sufficient to maintain the necessary glycolysis and there is a relative increase of the concentration of NADH together with a build-up of pyruvate. The changed ratio of $NAD^+$ to NADH makes the cofactor more reducing and so pyruvate is converted to lactate, making up for some of the $NAD^+$ deficiency but resulting in feelings of fatigue and pain:

$$\underset{\text{pyruvate}}{C_3H_3O_2^-} + NADH + H^+ \rightarrow \underset{\text{lactate}}{C_3H_5O_2^-} + NAD^+$$

The enzyme involved is lactate dehydrogenase. Remember that enzymes, like all catalysts, do not influence the outcome of a reaction; they affect only its speed. It is the same enzyme that catalyses both the forward and the reverse reactions.

When the excessive exercise is finished respiration is at a sufficient rate for the levels of $NAD^+$ and NADH to return to normal. The $NAD^+$ and NADH ratio re-establishes it as an oxidizing agent and the lactate is converted back to pyruvate.

## 4.4 Applications of potentiometric measurements

The two main uses of potentiometric measurements are in the measurement of concentrations and in titrations.

### 4.4.1 Measuring concentrations with ion-selective electrodes

The practical set-up for using ISEs is as illustrated in Figure 4.9, except that the platinum electrode would be replaced by an ISE.

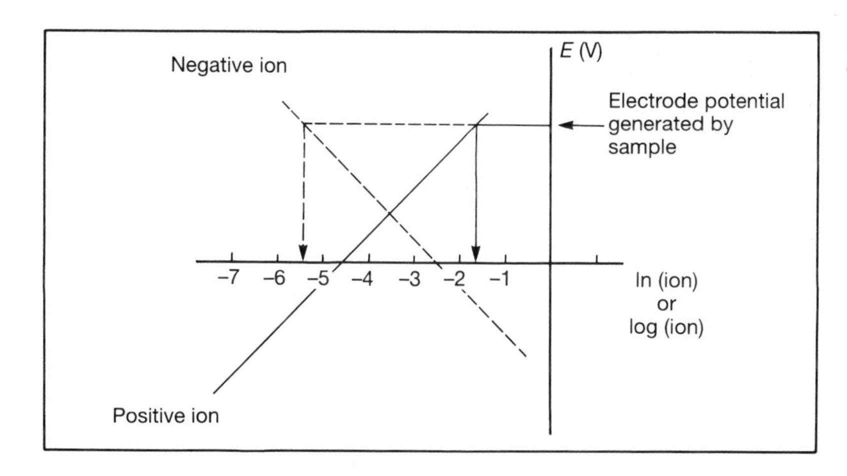


**Figure 4.10** Ion-selective electrode calibration graphs.

The potential developed by the ISE is determined by using the equation:

$$E = \text{emf} + E_{\text{ref}}$$

as was described in Section 4.3.

When we use ISEs the Nernst equation shows us that the value of the electrode potential is related to the logarithm of the concentration of the ion (ln[ion] or log[ion]). Thus, a graph of electrode potential against either ln[ion] or log[ion] is, theoretically, a straight line. It does not pass through the origin but should cut the $y$-axis where the electrode potential equals $E_0$. However, you are unlikely to be making measurements of concentrations in this region (near 1 mol dm$^{-3}$). The gradient of the graph is positive for positive ions and negative for negative ions (see Box 4.3 and Figure 4.10).

For calibration you prepare standard solutions whose concentrations form a logarithmic series (see serial dilution in Section 2.6). You then plot the electrode potential readings obtained against the logarithm of the concentration. You can use the graph as long as a straight line (or nearly straight) is obtained over a reasonable range of concentrations. Remember that the figure obtained from the calibration graph must be converted back from a logarithm to an ordinary number (using a function such as 'inv ln', sometimes labelled '$e^X$'; or 'inv log', $10^X$, on your calculator). Thus, using the values shown in Figure 4.10:

Theoretically the slope of the graph can be calculated from the *RT*/*nF* factor in the Nernst equation (see Box 4.3). In practice, the experimental slope may not be exactly in agreement and we would say the electrode was not responding ideally. This does not prevent it being used for determining concentrations.

for a positive ion such as sodium
if you used natural logarithms for the $x$-axis, $\ln[Na^+] = -1.7$, so $[Na^+] = 0.183$ mol dm$^{-3}$
but, if you used base 10 logarithms, $\log[Na^+] = -1.7$, so $[Na^+] = 0.020$ mol dm$^{-3}$
(or whatever units you used for the concentrations of the standards)

At this stage you should take into account any dilution of the test sample that you might have done. Do not multiply the logarithm value by a dilution factor.

### 4.4.2 Potentiometric titrations

*General redox titrations*

The experimental set-up is likely to be as shown in Figure 4.11. During the titration the reactant in the flask is converted to a product. If the reactant is a reducing agent the reagent oxidizes it and the concentration of the reduced form, [red], diminishes to

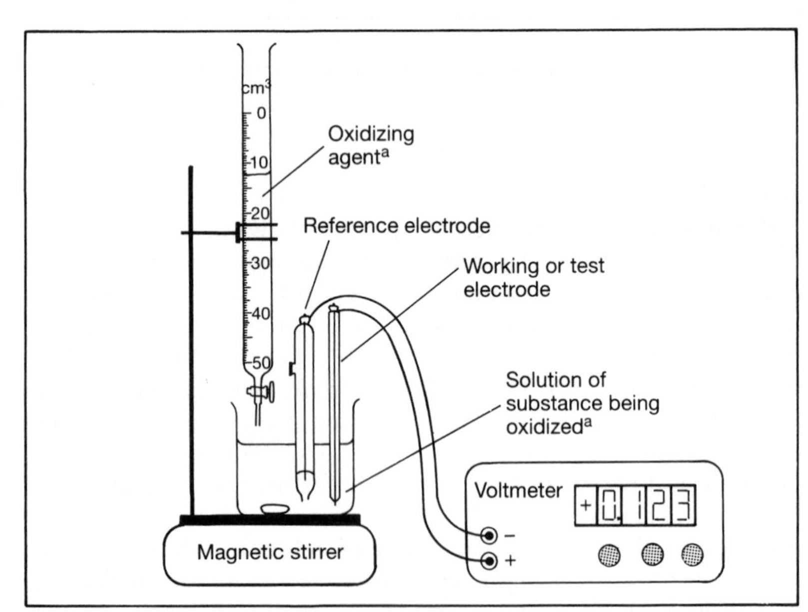


**Figure 4.11** A potentiometric titration. [a] Note: you can put a reducing agent in the burette, in which case the solution in the beaker undergoes reduction.

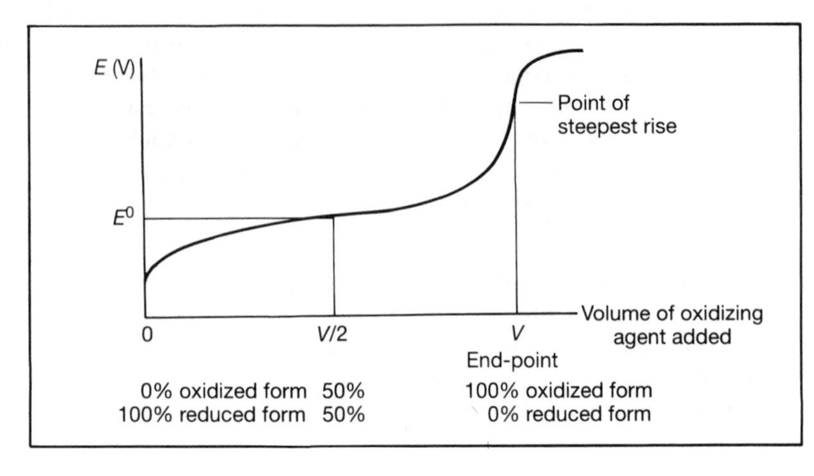


**Figure 4.12** A typical potentiometric titration curve.

(virtually) zero while that of its oxidized form, [ox], starts at (virtually) zero and increases. The opposite is the case if the reactant is a reducing agent. Note that the reagent being added from the burette does not contribute to the redox couple being measured, it merely causes the change from one form of the reactant to the other. In either case, the ratio of the oxidized form to the reduced form varies in such a way as to give the characteristic S-shape, shown in Figure 4.12, for a graph of the electrode potential against volume of reagent added.

The end-point of the titration corresponds to the point of steepest rise in the graph. After the end-point there is no further reaction and the shape of the graph is determined only by dilution effects. A special significance is attached to the point on the graph half-way between the start and the end-point. At this stage the concentrations of the oxidized and reduced forms are equal and their ratio is therefore 1, so the logarithmic term in the Nernst equation becomes ln 1, whose value is zero. Hence, at this point, the experimental $E$ value equals $E_0$ (and also approximately equals $E^{\ominus}$).

Remember that the logarithm of 1 to any base is zero. Try it on your calculator using the log and the ln buttons!

Such titrations may be used two ways.

1. You can find the end-points of titrations without having to use indicators, which avoids the sometimes difficult judging of colour changes. You can then do the usual sorts of titration calculations of concentrations.
2. You can determine $E_0$ values and establish the relative oxidizing strengths of particular systems.

*With ion-selective electrodes*

*The pH electrode.* This is, of course, an ISE since it responds only to changes in hydrogen ion concentration. Such an electrode is used by pH meters, which are able to display the pH directly.

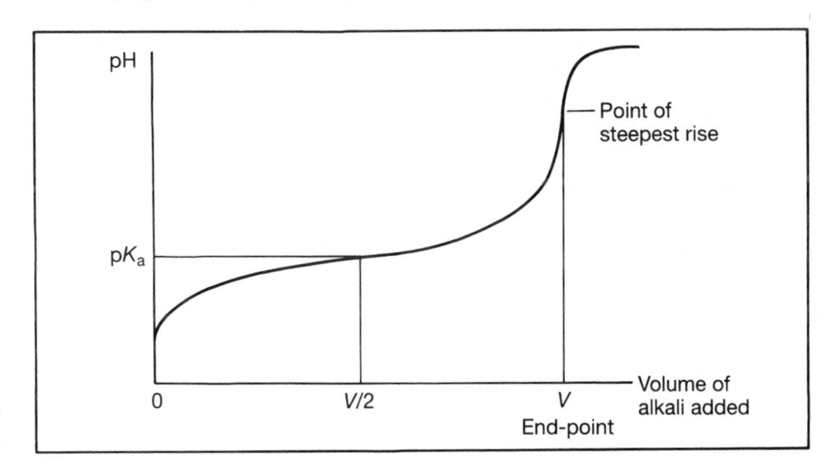


**Figure 4.13** A typical pH titration curve.

They actually measure the voltage developed by the electrode and are usually able to display this (as mV) as well as the pH. The conversion from mV to pH is simple since both are logarithmically related to the hydrogen ion concentration (see Box 4.4). pH titration curves have the same shape as redox titration curves because the same logarithmic form of equation applies (see Box 4.1). We can therefore find the end-points of acid–base titrations in the same way as for redox titrations and the $pK_a$ (see Box 4.1 and Figure 4.13) of weak acids in the same way as for $E_0$.

**Box 4.4 Handling glass electrodes and calibrating a pH meter**

As you might expect, glass electrodes are quite fragile; you must handle them with care. It is important that the sensitive glass bulb is kept moist to retain its activity. Electrodes are usually stored vertically with the bulb immersed in 0.1 mol $dm^{-3}$ HCl or 1 mol $dm^{-3}$ KCl. During use they should be kept in solutions as much as possible, and rinsed well with deionized water in between measurements. If any adhering water might affect the next pH measurement you should gently blot the bulb with a tissue so that distinct drops of water are removed. If the electrode has become dried out, prolonged soaking in 0.1 mol $dm^{-3}$ HCl may restore it, but this cannot be guaranteed. A malfunctioning electrode tends to give sluggish response, and to require abnormal adjustments to the meter.

A graph of the emf produced by a pH electrode is a straight line since both pH and mV are logarithmically related to the hydrogen ion concentration. When you set up a pH electrode and meter the response is unlikely to follow the ideal line exactly because of 'junction potentials' and non-ideal behaviour of the glass membrane, so you must calibrate the meter before you take any readings of samples.

Calibration is done in three steps – allow about 5 min after switching on for stabilization.

1. Adjust for temperature: use the appropriate knob or key sequence. pH changes with temperature. For much biological work you will be working at constant temperature so this effect will not be noticeable, but the important point is that calibration and measurement must be done at the same temperature. Some meters allow you to connect a temperature probe so that automatic temperature adjustment can be made.
2. Set pH 7: place the electrode in a standard pH 7 buffer. Use the knob labelled 'set buffer' or some similar phrase, or use the appropriate key sequence, to bring the display to pH 7. The design of a pH electrode is such that it gives 0 mV with respect to its reference electrode at pH 7, so this is like setting the zero, which is necessary in most electronic meters.
3. Adjust the 'slope': place the electrode in another standard buffer (usually pH 4 for working with acidic solutions and pH 9 for alkaline solutions) and use the knob labelled 'slope' or the appropriate key sequence to bring the display to the pH value of the buffer. This controls the sensitivity of the meter, which is equivalent to changing the slope of the calibration graph.

Some meters these days require only a 'one-point' calibration. The meter can also be conveniently checked by an electronic tester, which generates the potential that an electrode would give at various pH values. This does not, of course, check the electrode but it does allow you to distinguish between meter faults and electrode faults. The former tend to cause rapid and erratic changes in the reading, while the latter are usually slower and more uniform (such as drift in the readings).

If the meter is kept switched on the calibration will usually last for the whole day.

*Chloride titration.* You can follow the progress of a titration of a chloride solution against silver nitrate using a chloride electrode. The combination of the silver ions with the chloride ions to give insoluble silver chloride effectively removes the chloride. Again, a typical titration curve has the same shape as those in Figures 4.12 and 4.13. Thus you can detect end-points and calculate concentrations of solutions in the usual manner for titrations.

## 4.5 Amperometry and related techniques

### 4.5.1 Amperometric titrations

In the previous sections we were concerned with the potentials developed by electrodes, and noted that no measurable current should flow for those measurements. In contrast, when we use the technique of amperometry we are applying a voltage and measuring the resulting current. This technique, then, has some similarities to conductimetry and to electrolysis. On the one hand the current flowing is proportional to the concentration of the ions under controlled conditions while, on the other hand, the products of the electrolysis can be used to determine these concentrations.

The electrodes you use will usually be inert electrodes but, although they do not participate in the electrode reactions, they do become polarized by the products of the electrolysis, and the set-up usually has some form of stirring or cleaning of the electrode surface. The amount of electrolysis must be very small compared with the amount of substance being analysed so as to avoid changing the concentration that you are trying to measure! yet another precaution is to use a voltage that will not indiscriminately discharge other ions in solution. The following titration is typical.

*The amperometric titration of vitamin C*

Vitamin C is readily oxidized ($E_0'$ + 0.06 V; see Table 4.1) and may be titrated with the mild oxidizing agent, 2,6-dichlorophenolindophenol (DPIP; $E_0'$ + 0.26 V; see Table 4.1). This reagent is a deep purple colour but its reduction product is virtually colourless, so you can make a visual estimation of the end-point (i.e. when a permanent pink colour just appears). However, if you had to estimate the vitamin C content of blackcurrant juice you would find it impossible to detect the end-point, so alternatives methods have to be used.

For the amperometric method you place a pair of inert electrodes in the solution and set at suitable voltage (slightly greater than

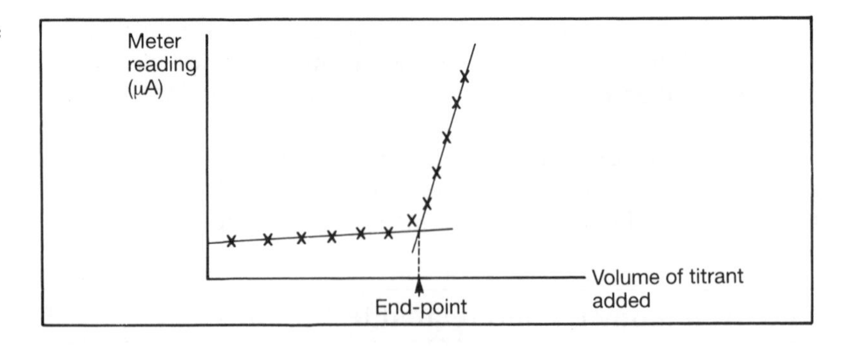


**Figure 4.14** A typical amperometric titration curve.

+0.26 V). This will reduce the DPIP by supplying electrons at the negative electrode. You can then detect the presence of the DPIP reagent by a sensitive ammeter. Before the end-point of the titration the DPIP reacts with the vitamin C before it can be detected by the electrodes. After the end-point there is no vitamin C left to react and the excess DPIP is detected. The change from no current flowing to some current flowing marks the end-point. The usual experimental procedure is to add aliquots of the DPIP from a burette, noting the ammeter reading until the total volume added is well past the end-point. A graph of ammeter reading against the volume of DPIP added shows two distinct straight portions, and the end-point is at their intersection (Figure 4.14).

You first do a titration with a known concentration of vitamin C and then with the test solutions. The concentrations of the test are then simply calculated by proportion thus:

Suppose 5 $cm^3$ of a standard solution of ascorbic acid (pure vitamin C) containing 1 mg $cm^{-3}$ requires 3.0 $cm^3$ of the DPIP. Then 1 $cm^3$ of the reagent is equivalent to 1/3.0 = 0.333 mg ascorbic acid.

5 $cm^3$ of a test solution requiring, say, 3.2 $cm^3$ of the reagent must contain 3.2 × 0.333 = 1.066 mg. Therefore the concentration in the test solution is 1.066/5 = 0.213 mg $cm^{-3}$.

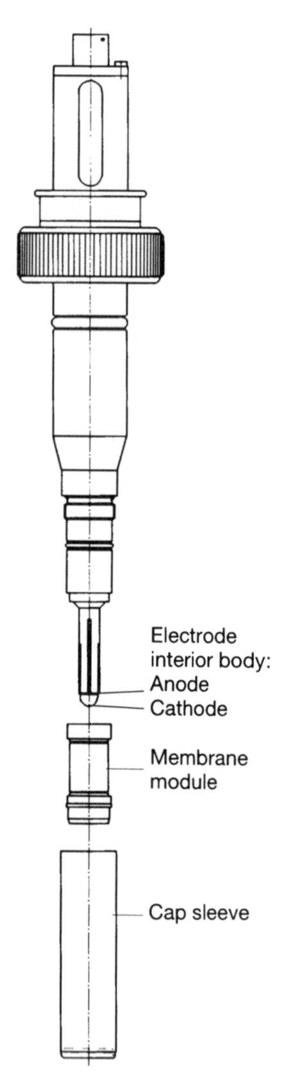


**Figure 4.15** The construction of a Clark-type oxygen electrode. (Courtesy Mettler-Toledo Ltd.)

## 4.5.2 The oxygen electrode

Oxygen in solution can be reduced electrochemically using a platinum microelectrode maintained at −0.6 V with respect to an Ag/AgCl reference electrode. The platinum electrode, being negatively charged (the cathode), provides electrons for the reduction process, and the cathode reaction is:

$$O_2 + 2H^+ + 2e^- \rightarrow H_2O_2 \quad (E'_0 = 0.3\,V;\ \text{see Table 4.1})$$

Reductions of other molecules or ions in the solution might also occur, so the cathode is protected from the bulk of the test solution by a thin membrane through which only the oxygen diffuses

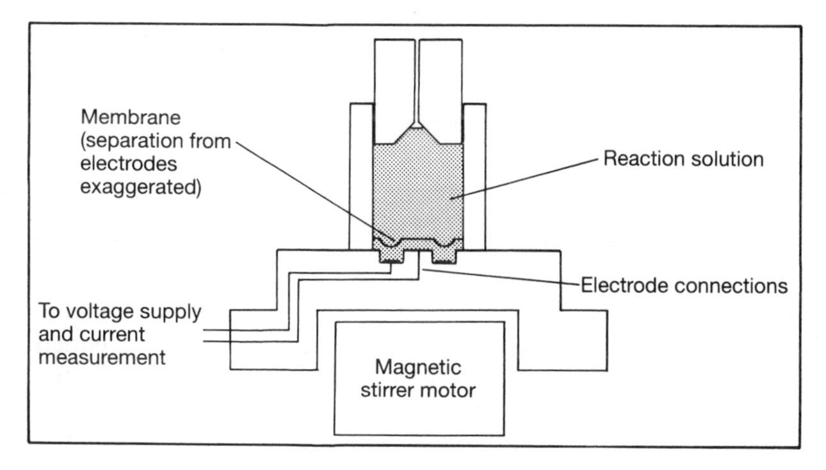


**Figure 4.16** The Rank oxygen electrode assembly. (Redrawn from Rank Bros, UK.)

(usually 12 mm 'Teflon', although cellophane, 'clingfilm' and polythene are other possibilities). The current produced by the reduction is proportional to the dissolved oxygen concentration.

There are two main styles of oxygen electrode.

1. The **Clark electrode** is a probe type and can be used in the same sort of way as a pH electrode. This style is particularly suitable for, say, the monitoring of oxygen levels in lakes and rivers.
2. The **Rank electrode** is not portable, but can follow changes in oxygen levels and is suitable for the study of the kinetics of enzymes such as glucose oxidase – the effect of pH, say, on the activity of the enzyme. It may also be used for the quantitative analysis of substrates of enzymes; the specificity of the enzyme allows you to determine glucose in serum or in fruit juices even though other sugars may be present.

In principle, you can use an oxygen electrode to monitor any reaction which involves the use or evolution of oxygen.

### 4.5.3 The chloride meter

Chloride measurements are often required in biological work, particularly in biomedical assays. Although we have described a chloride electrode (which uses potentiometric measurements) there is a coulometric alternative. A small accurately measured volume (0.5 $cm^3$) of a sample, such as brine from food processing or serum from a blood sample, is pipetted into a specially formulated acid solution in the apparatus. A steady current is passed between a pair of silver electrodes immersed in the sample and silver ions are released into the solution at a constant rate. They combine with the chloride ions to give insoluble silver chloride. When all the chloride in the sample has been precipitated there is a change in the conductivity of the solution. The meter detects this change and is calibrated by using standards to give a read-out directly in $mg\,dm^{-3}$. The system is less sensitive than a chloride electrode but it is easier to operate and to calibrate.

## 4.6 Polarography

We will introduce the principle of polarography here and indicate some of the many derivatives of the technique. The experimental set-up is rather elaborate, but in recent years it has been refined and its operation has been made much simpler, mainly due to the advent of microprocessor control. So it is now a technique that can claim to be routine, although you need some time to gain the necessary practical experience to be able to run it reliably.

In polarography both the voltage and the current are varied in a circuit containing electrodes. In the section on potentiometry we noted that $E^{\ominus}$ measures the tendency of a half-reaction to gain electrons and act as an oxidizing agent. For an ideal electrode (where the half-reaction was completely reversible), if we apply a potential more positive than the $E^{\ominus}$ no reaction occurs, but if we apply a more negative potential then the half-reaction accepts electrons and a current does flow in order to supply them. We can also use the technique in the reverse sense and induce oxidation in suitable half-reactions by applying a sufficiently positive potential. In practice, the electrodes have a marked influence on the potential required to bring about reduction or oxidation but this does not affect the general working of the technique.

### 4.6.1 Simple direct current polarography

Here, the applied voltage is steadily varied and the current flowing through the solution is measured. The results are plotted as a graph of current against voltage: a **polarogram**. The commonest form of polarography uses an increasingly negative voltage and the polarogram has the outline shape shown in Figure 4.17.

At the beginning very little current flows because the voltage is not negative enough to reduce the substance in the solution. This baseline is known as the **residual current**. As the voltage is increased there comes a point where there is a distinct increase in the current, and a step (or wave) in the polarogram. The potential at the half-way point of the wave (the '**half-wave potential**') is characteristic of the half-reaction occurring and can help in identifying the species involved. As the potential increases still further, the polarogram levels off to give a **limiting current**. At this stage the ions are moving through the solution at a rate limited by their ability to diffuse through the solution. This is

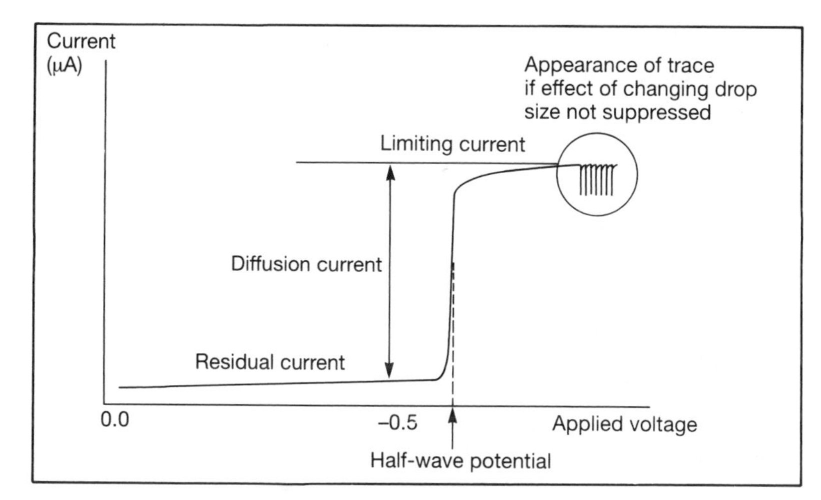


**Figure 4.17** A typical simple DC polarogram.

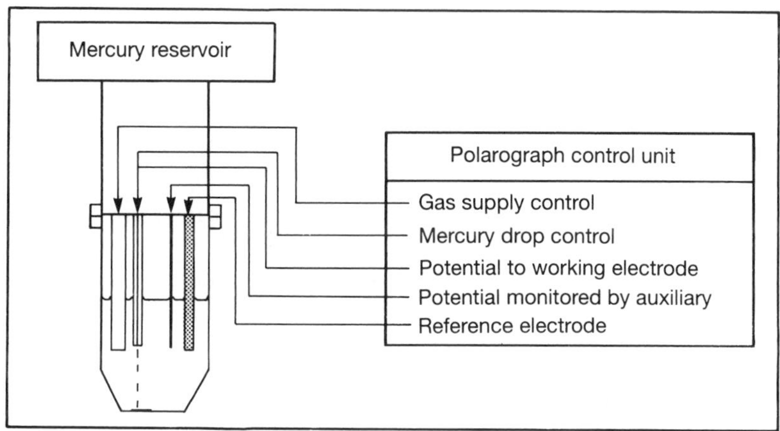


**Figure 4.18** A typical simple DC polarographic cell.

related to their concentration, and hence the wave height (known as the **diffusion current**) is a measure of concentration. Calibration of wave heights against concentration gives a means of determining unknown concentrations.

Polarography has the advantage of requiring little sample preparation, and can discriminate between different ions in the same solution. However, there are several practical points that must be observed in order to get results as simple as described above.

- The electrode surface must be kept fresh so that it is not polarized by electrolysis products. This is usually done by having mercury dropping from a very fine capillary tube ('DME' = dropping mercury electrode). The formation and detachment of the drop ensures that there is always some fresh electrode surface. However, the changing size of the drop produces a zig-zag effect in the polarogram (see enlargement in Figure 4.17). If this is too severe, it can be suppressed electronically.

- Some solutions produce a peak in the polarogram just before the limiting current is reached. The reasons for this are not entirely clear but it is found that such peaks can be suppressed by adding a trace of non-ionic detergent to the sample.
- Oxygen is easily reduced, in the same way as described for the oxygen electrode. It is present in most solutions that have been exposed to the air. It interferes when negative potentials are used and must be removed. The usual method is to bubble nitrogen through the solution before the polarograph is started, and ensure that a layer of nitrogen shields the top of the solution from absorbing more oxygen from the air while the experiment proceeds.
- Because the working electrode is delivering a current, it does not register the applied potential reliably, so an auxiliary electrode is used for monitoring the true potential.

### 4.6.2 Derivatives of the technique

By today's standards, simple direct current polarography does not achieve particularly low detection limits. Refinements of the technique such as the use of a momentarily static hanging drop, or a pulsed application of the potential (as in **differential pulse polarography**), achieve much lower limits and the technique rivals other methods of trace analysis such as in pollution by heavy metals.

### 4.6.3 Related methods

Further modifications of the basic technique have also been developed. The potential may be increased and then decreased as in **cyclic voltammetry**. Alternatively, a potential may be applied for a time to allow the accumulation of the analyte in the mercury, followed by its release on reversing the potential; this is known as **stripping voltammetry**.

## Further reading

For a fuller treatment of the theory and the instrumentation you can consult:

Braun, R.D. (1987) *Introduction to Instrumental Analysis*, McGraw-Hill International Edition, ISBM 0 07 100147 6.
Chapters 22 and 23.

Harris, D.C. (1987) *Quantitative Chemical Analysis*, 2nd edn, W.H. Freeman, ISBN 0 716 71817 0.
Chapters 10–12, theory of acids and bases; Chapters 15–17,

redox theory and potentiometric titrations; Chapter 19, polarography and amperometry.

Skoog, D.A. and West, D.M. (1980) *Principles of Instrumental Analysis*, 2nd edn, Holt & Saunders International Edition, ISBN 4 8337 0003 4.

Chapters 18–21.

Hart, J.P. (1990) *Electroanalysis of Biologically Important Compounds*, Ellis Horwood, ISBN 0 13 252107 5.

If you need to develop your knowledge of the various types of polarography and voltammetry you will find much detailed information here. Note, however, that potentiometry and ion-selective electrodes are not included in this title.

Morris, J.G. (1974) *A Biologist's Physical Chemistry*, 2nd edn, Edward Arnold, ISBN 0 7131 2414 8.

An extremely thorough and comprehensive text. If you have a fair grasp of the principles of electrochemistry this will help you consolidate your understanding. It also deals with thermodynamics and kinetics.

The ACOL 'teach yourself' style books include the following titles.

Evans, A. (1987) *Potentiometry and Ion Selective Electrodes*, Wiley, ISBN 0 471 91393 6.

Riley, T. (1987) *Polarography and Other Voltammetric Methods*, Wiley, ISBN 0 471 91395 2.

Riley, T. and Tomlinson, C. (1987) *Principles of Electroanalytical Methods*, Wiley, ISBN 0 471 91330 8.

# 5 Chromatographic techniques

Even if you have not done a chromatography experiment as such, you may well have seen the results of the process. When a solvent such as alcohol is spilled onto writing paper with black ball-point ink on it, colours develop as the liquid spreads across the ink marks. This is because the ink contains several pigments which altogether look black but, when separated by the action of the flowing solvent, show their individual colours.

The discovery of this separation technique, now known as chromatography, has been described in many books and papers. The name chromatography was devised because the effect was originally observed with coloured substances – chlorophyll and other pigments extracted from plants. Today, the term is used whether the substances are coloured or not. Separation techniques are particularly important for biological analyses since many biological samples are complex mixtures. Chromatography does not guarantee to separate all the components of a mixture completely but it leads the way into both qualitative and quantitative analysis. Additionally, the separation is non-destructive and it is possible to use the separated components of your sample for further work. Yet again, you can scale-up the process so that a purified sample may be isolated.

The appeal of chromatography is that, although it is a powerful analytical method, it can be done using quite simple apparatus; a piece of paper or a coated glass plate in a glass tank, or a narrow column filled with adsorbent material. Of course, it has been elaborated into various sophisticated instrumental techniques that are now routinely used in all sorts of laboratories all around the world. By the way, there is now a gas chromatograph even on Mars!

After working through this chapter you should:

- be acquainted with chromatography systems;
- be thoroughly familiar with the nature of stationary and mobile phases;

- appreciate practical aspects of the technique;
- be able to interpret simple chromatograms qualitatively and quantitatively;
- recognize the basic features of more complex instruments;
- have an appreciation of the mechanisms involved in chromatography.

## 5.1 Survey of chromatographic techniques

### 5.1.1 Simple chromatography

There are two main classes of chromatography: column and planar. The way in which they produce separations is, of course, the same but the way in which they are handled is somewhat different. However, the differences are more apparent than real, and you should look for parallels between the two in what follows.

*Column chromatography*

Historically this was the first method to be developed. Typically the column may be 10–50 cm high and 1–2 cm wide although much smaller columns – 'microcolumns' – are used for some nucleic acid separations. It contains an adsorbent material in the form of a fine powder which is known as the **stationary phase**.

Adsorbent means attracting substances to the surface of the material.

You put a small volume (0.5–5 $cm^3$) of sample on the top of the column, and then start a flow of solvent, the **mobile phase**, through the column. The solvent emerging at the bottom is called the **eluate** and is usually collected in portions of 1–2 $cm^3$. The components in the sample are moved by the mobile phase at different rates so their time of arrival (the **retention time**, $t_R$) is different and they are eluted (i.e. emerge from the column) in different portions ('fractions') of the mobile phase. We say that their **elution volumes** ($V_e$) are different. If your sample contains coloured substances you can see which fractions contain components from the sample, but for colourless substances some more elaborate method of detection, such as ultraviolet (UV) absorption, has to be used.

You may be able to identify a component in the sample by its retention time or its elution volume. If it is the same as that of a known substance (a standard) run under identical conditions then that component may be the same as the standard. You cannot be sure unless you do a lot more chromatography or other tests and you should only use standards that you have good reason to believe are in the sample. The relative amounts of the components in the fractions can be judged from the intensity of the colours (if any) or by the readings from the detector.

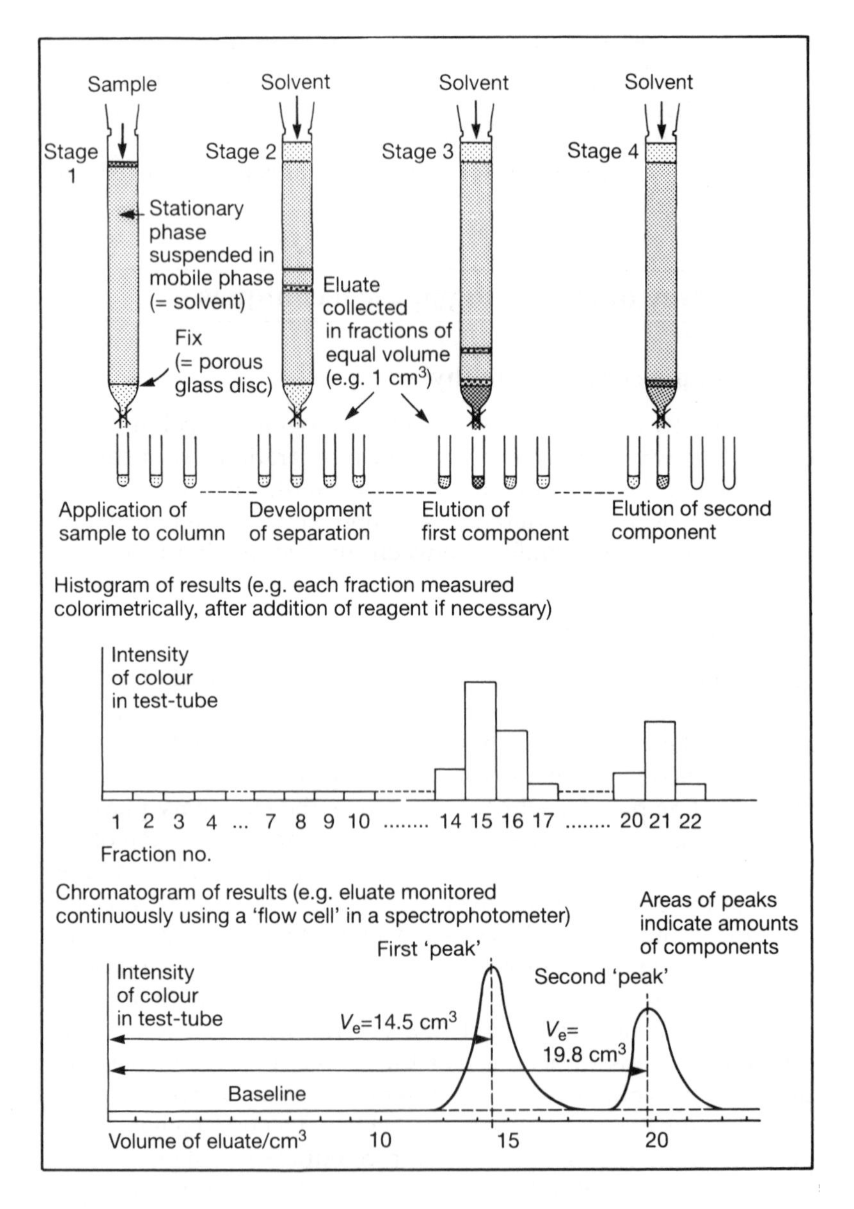


**Figure 5.1** Running a column chromatogram.

When all the fractions have been completely eluted, the column is ready for another sample. However, crude samples often leave strongly adsorbed (and usually highly coloured) material near the top of the column. This may affect subsequent samples and the number of times a column can be used may be limited.

*Planar chromatography; paper and thin-layer chromatography*

As an alternative to the column, you can use paper as a stationary phase (paper chromatography, PC). You can also use a thin layer

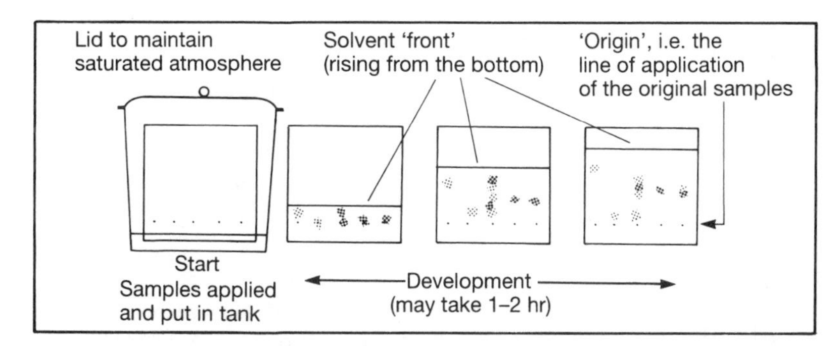


**Figure 5.2** Development of a planar chromatogram. Note: if the spots are colourless you will have to spray the finished chromatogram with a reagent to make them visible.

of adsorbent material (silica or alumina as in column chromatography) spread onto a glass plate, a plastic film or an aluminium sheet. This is thin-layer chromatography (TLC); it can give better results because there is less diffusion than with paper chromatography.

You put small spots of the test samples near one edge of the paper or plate, along with spots of standards, as shown in Figure 5.2. The paper or plate is then usually stood in a small quantity of the mobile phase in a glass tank with a lid to maintain a saturated atmosphere. This method is known as ascending chromatography since the mobile phase rises up the paper or plate by capillary attraction. Alternatively, paper may be hung from a trough which contains the mobile phase (descending chromatography). The position where each spot is applied is known as the **origin** and the moving boundary of the solvent is known as the **solvent front**. The position the solvent front reaches by the end of the chromatography is usually 1–2 cm from the edge of the chromatogram opposite the origin.

As with column chromatography, you identify components in a sample by comparing them with the standards run on the same plate or paper. In order to make proper comparisons, and to make allowances for variations between runs, it is usual to record results by means of the **retardation factor ($R_f$)**, which is calculated thus:

$$R_f = \frac{\text{distance from origin to spot centre}}{\text{distance from origin to the solvent front}} = x/y \text{ in Figure 5.3}$$

The range of values for $R_f$ can only be between zero and one. You cannot measure the distances $x$ and $y$ very precisely and so you should quote the $R_f$ values to no more than two decimal places. When the spots are irregular in shape you measure to the densest point of the colour. If the $R_f$ value is close to zero or close to one (near the origin or near the solvent front) proper chromatography has not taken place and identifications will not be reliable. The separations and identifications are most sensitive when the $R_f$ is in the range 0.1–0.9.

**Figure 5.3** Interpreting the results of a planar chromatogram.

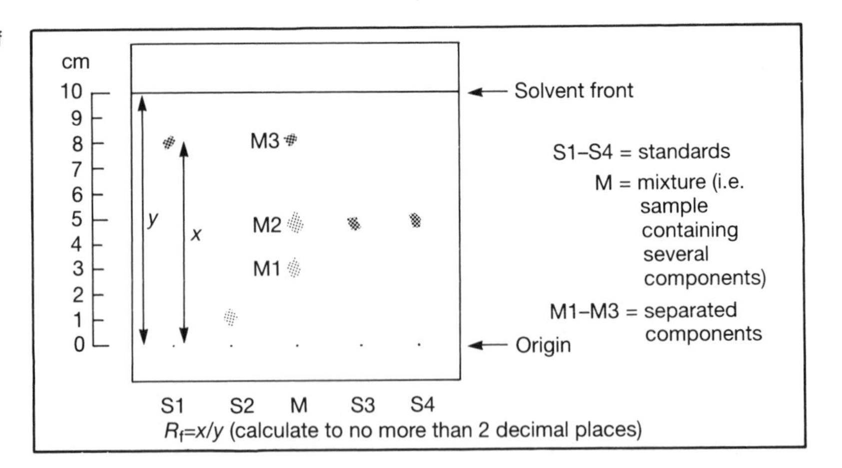


The interpretation of this chromatogram is that the test mixture contains three components: M1, which is different from any of the standards and is thus not yet identified; M2, which could be either S3 or S4 or both; and M3, which is most likely to be S1.

*Methods of detection for planar chromatography.* If your sample contains colourless components you will not be able to see where they are! You then have to use one of the following methods for locating the spots.

- You can spray plate or paper with a suitable reagent (and heat it if necessary) to produce colours.
- You can use UV absorption. The stationary phase has to have a fluorescent material in it. Components that absorb UV show as a dark patches on a bright background when the chromatogram is illuminated by a UV lamp. (You should not look directly at the UV lamp.)
- A densitometer is an adaptation of a UV–visible spectrophotometer which measures absorption or reflectance along the track of the sample. It gives a graph of the intensity of absorption against distance along the chromatogram.
- You can also do a similar sort of scanning with a Geiger counter assembly for radioactive samples. The results from a scan look like the chromatogram shown at the bottom of Figure 5.1.

*Methods of quantifying the results from planar chromatography.* As with all methods of analysis, it is just as important to obtain quantitative results as qualitative. Here are two procedures.

- Cut out the spot from a paper, or scrape off the area of a spot from a thin layer plate and put the material into a suitable solvent. You can then measure the amount of the component using one of the methods described in the chapters on spectrophotometry and electrochemistry.
- You can measure the areas of the peaks (see Section 5.3 for ways of doing this) in a densitometer or radioactive scan.

**Box 5.1 Practical tips for planar chromatography**

**Do:**

- make a note of the spots you apply to the chromatogram;
- have a saturated atmosphere in the tank;
- keep spots away from the edges of the paper or plate;
- keep spots as compact as possible;
- mark the position of the solvent front when you remove the chromatogram from the tank.

**Don't:**

- put spots too close together along the origin;
- draw lines through the layer of a TLC plate;
- use a pen to mark a paper chromatogram (why not?);
- have the origin line immersed in the solvent;
- calculate $R_f$ values to more than two decimal places.

## 5.1.2 Extending the scope of simple chromatography

Simple chromatography can produce some remarkable separations but many biological samples offer a great challenge because of their complexity. Various improvements or refinements have been used to tackle some of the problems without requiring expensive instruments. Some are noted here to give an indication of the capabilities of simple chromatography.

*Two-dimensional chromatography*

For complex mixtures (such as the amino acids from a hydrolysed protein) one solvent mixture may not be sufficient to separate all the components. In such a case, you can use two-dimensional

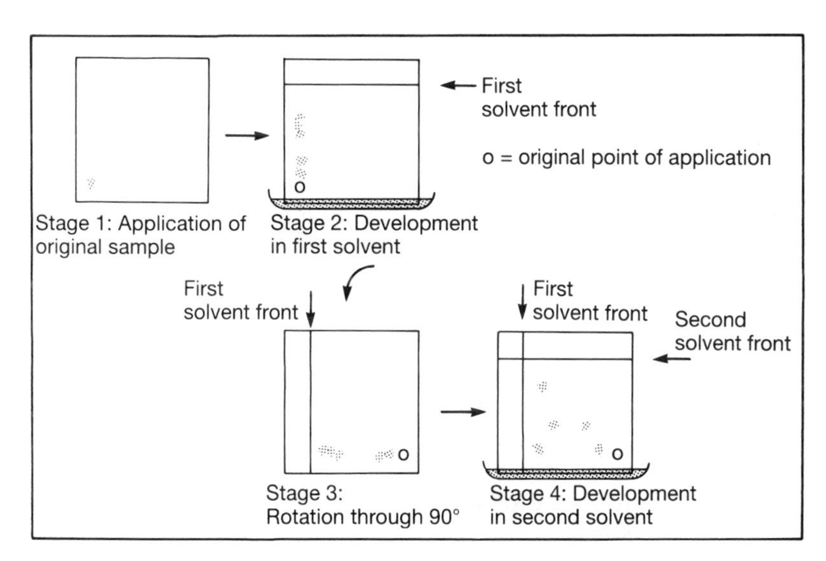


**Figure 5.4** Two-dimensional chromatography.

A similar procedure can be used with electrophoresis being used in one dimension and chromatography in the other.

(2D) planar chromatography. A single sample is applied to one corner of the chromatogram, which is inserted into the tank so that it develops normally in one direction. After that, the chromatogram is dried, turned through 90° and placed in a second (different) solvent mixture and developed again. The components are now spread out over an area, with a much greater chance of separation.

*Silver nitrate loaded stationary phase*

It has been found that compounds containing C=C bonds (especially the unsaturated fats and oils) can be separated much more effectively in TLC by using plates spread with silica containing some added silver nitrate than by using silica alone. The double bonds in the unsaturated fats form a molecular attraction for the silver ions which is stronger if there are more double bonds in the molecule. So separations are achieved on the basis of varying degrees of unsaturation. However, the method has probably been superseded by gas chromatography.

*Multiple passes of solvent*

For spots with small $R_f$ values the distance moved can be exaggerated by rerunning the initial chromatogram several times. Faster moving spots are less affected by the procedure since they are not immersed in the rerun solvent for so long. The technique was first used with multiple passes of the same solvent, but has now been extended to a changing solvent system.

*Grooved plates*

Where crude samples have to be analysed for trace components, TLC plates with a groove cut at the position of the origin can accommodate larger samples. If most of the crude sample is strongly retained (it stays at the origin) as is usually the case, then the small quantities of components of interest are eluted into the rest of the area of the plate where normal chromatography takes place.

*Chiral stationary phases*

Mixtures of optically active isomers ('enantiomers'; you may have to look this up in an organic chemistry or a biochemistry textbook) cannot be separated on normal stationary phases but chiral stationary phases can produce separations and are now widely used, especially in the pharmaceutical industry where the chiral purity of drugs is important.

*Preparative-scale chromatography*

Chromatography is principally an analytical method and so uses small quantities. The separation process is non-destructive and, although some of the detection methods are destructive, others are not. If you use one of these you should be able to recover the separated components from the collected fractions of the eluate. You can expect these components to be pure, so chromatography can be a sort of purification process. If you scale it up to deal with larger samples you may be able to isolate components in quantities up to 1 g or so, although you may not obtain such good separation as you get for the analytical work. For substances which may be difficult or impossible to purify by other means this can be an important procedure.

*Flash column chromatography*

In order to increase the separation obtained in column chromatography, it is possible to use more uniform stationary phase materials together with a small applied pressure. This is claimed to produce analytical-quality separations with larger samples.

## 5.2 How chromatography works

The essential components in chromatography are:

1. the **stationary phase**, i.e. the material in the column, or the layer on the plate, or the paper sheet;
2. the **mobile phase**, which is the solvent (often a mixture) that percolates through the stationary phase. (If you are unsure what a 'phase' is, see Box 5.2.)

**Box 5.2 Phases**

A phase may be a solid, a liquid or a gas. It is normally considered to be uniform in composition and contained within definite boundaries (interfaces). An example of a two-phase system is a partly filled bottle of water. Inside the bottle there is a liquid phase – the water – and a gas phase – the air and water vapour mixture. You might ask whether the material of the bottle could be regarded as a third phase. Strictly the answer is yes but normally the container is assumed not to interact in any way with its contents; it is said to be inert. It does not influence the behaviour of the phases and is not considered to be part of the system. Other examples include the contents of an unfertilized egg (if we ignore the membranes there are two liquid phases – the yolk and the white) or the bed of a river (one solid – the particles of the silt – and one liquid – the water).

How many phases are there in a partly filled bottle of ordinary milk? (Think carefully about this!)

**Table 5.1** Possible combinations of phases in chromatography

| *Stationary phase* | *Mobile phase* | |
|---|---|---|
| | *Liquid* | *Gas* |
| Liquid | Counter-current distribution (CCD) | Gas–liquid chromatography (GLC)[a] |
| Solid | Paper chromatography (PC)<br>Thin-layer chromatography (TLC)<br>High-performance thin-layer chromatography (HPTLC)<br>Liquid chromatography (LC)<br>High-performance liquid chromatography (HPLC) | Gas–solid chromatography (GSC)[a] |

[a] Often considered together as gas chromatography (GC).

Since chromatography requires one phase to be moving, and the other stationary, the number of possible combinations of solid, liquid and gas which can give rise to a chromatographic system is restricted to four (see Table 5.1).

### 5.2.1 Distribution of solutes between phases

Before describing the processes involved when a chromatogram is run, let us consider what happens to a single substance introduced into a stationary two-phase system. We need to assume that this substance is in some sense soluble in both phases and we will refer to it as the **solute**.

With a solid phase the solute will adhere to its surface. This is a process known as **adsorption**. When a phase is liquid there will be an ordinary solution in which the molecules of the solute are attracted to those of the solvent. For a gas phase there there will be a mixture of the solute and the gas, as long as the solute has a sufficient vapour pressure (it must be volatile). The extent of the adsorption, solution and vaporization is largely dependent on the **polarities** of the molecules (for more discussion of polarity, see Box 5.3).

**Box 5.3 The concept of polarity**

There is a marked difference in volatility between ethane, ethanol and water, as is shown by their boiling points (−89, +78 and 100°C respectively), even though they are molecules of similar mass. This is because some covalent bonds have an uneven sharing of electrons which leads to a slight positive charge at one end of the bond and a slight negative charge at the other. You find this when atoms of high electronegativity (O, N are the two most important ones) form bonds with atoms of lower electronegativity (particularly C, H and Cl). Such bonds are called 'polar' and they have various strengths. Polar bonds in a molecule produce attractions between positive parts of one molecule and the negative parts of others. This make it more difficult to vaporize polar substances, hence the higher boiling points. Another effect is that polar solutes dissolve more readily in polar solvents.

Some common bonds, arranged in order of decreasing polarity, are:

(All are neg.–pos.)

$$\text{O—H} > \text{O—C} \sim \text{N—H} > \text{N—C} \gg \text{Cl—C} > \text{C—H} > \text{C—C}$$

strongly polar ←——→ moderately polar ←——→ essentially 'non-polar'
less volatile ←————————————————→ more volatile
more soluble ←— (in polar solvents, e.g. water, ethanol) —→ less soluble
less soluble ←— (in non-polar solvents, e.g. petrol) —→ more soluble

Look again at the boiling points given for ethane, ethanol and water and account for them in terms of their polarities.

In compounds containing the bonds O—H and N—H, there is an additional attraction between the molecules known as **hydrogen bonding**. This enhances the effects of polarity. For a fuller treatment of electronegativity, polarity and hydrogen bonding you should refer to a textbook of organic chemistry.

For two-phase systems a solute will become 'distributed' between the two phases. This is because the solute molecules diffuse and pass across the interface between the phases in both directions (exchange). The concentration of the solute in each of the phases depends on the relative attractions between the molecules of the solute and each of the two phases. Different solutes will, of course, have different distributions.

The ratio of these concentrations is known as the distribution coefficient ($K_d$). It is also known as the partition coefficient when both phases are liquid. $K_d$ can be applied provided that neither phase is saturated. In chromatography saturation of one of the phases has adverse effects on the quality of results.

We can also use the distribution as the basis of an extraction procedure known as **liquid–liquid extraction**, where two solvents are shaken together so that the solutes become distributed; the two phases are then separated and each is shaken with fresh solvent, which produces further extraction. Separation of these phases and pooling of the similar ones enhances the separation of the solutes. When the procedure is repeated 10 or more times, say, in a specially designed assembly of tubes the technique becomes **counter-current distribution**. Both techniques can be used to separate (extract) components from mixtures and are related to chromatography but not generally classed as chromatographic techniques.

### 5.2.2 Movement in a chromatography system

In chromatography, the flow of the mobile phase past the stationary phase allows the distribution process to take place many times, and the solute molecules spend some of their time being moved by the mobile phase and some being held still by the stationary phase. The net effect is that a solute moves at an average rate less than that of the mobile phase. The actual rate will depend on the relative attraction (or affinity) of each phase for that solute. When there is more than one solute their rates of

**Table 5.2** Typical materials for chromatography

| *Stationary phase ('normal phase')*[a] | | *Mobile phase (this is a very abbreviated list)* | | *Stationary phase ('reverse phase'*[a] *mostly used in HPLC and HPTLC)* |
|---|---|---|---|---|
| $SiO_2$ | ↓ | Hexane | ↑ | C-8-silyl- |
| $Al_2O_3$ | | Benzene | | C-18-silyl- |
| 'Kieselguhr' | | Chloroform | | Cyanopropyl-silyl- |
| Cellulose | | Tetrahydrofuran | | etc. |
| | | Acetonitrile | | (see Section 5.5) |
| | | Butanol | | |
| | | Ethanol | | |
| | | Methanol | | |
| | | Water | | |

**Eulotropic strength**

The arrows indicate increasing power to move solutes along the column, plate or paper. It is mainly related to the polarity of the solvent (see Box 5.3 for more explanation of polarity).

When a single solvent will not produce the desired separation, mixtures of two, three or even four solvents may be used.

[a] See Section 5.1.

movement will be different and it is this that leads to separation.

In **normal-phase** chromatography (which might more properly be called 'normal polarity') the stationary phase is more polar than the mobile phase. The greater the polarity of the mobile phase the more the solute molecules are displaced from the stationary phase and the faster they move. We refer to this as increasing **eulotropic strength**. In normal-phase chromatography it requires increased polarity. However, in **reverse-phase** (also called reversed-phase) chromatography the stationary phase is less polar than the mobile phase and the eulotropic strength of a solvent is inversely related to its polarity (see Table 5.2).

In normal-phase chromatography the stationary phase is most frequently the polar substance silica. This has a strong affinity for polar solutes such as carboxylic acids so there is little movement. In contrast, non-polar solutes such as the triglycerides found in fats and oils will move almost as fast as the solvent because of the lack of affinity between them and the silica. In gas chromatography the mobile phase does not really have an affinity for the solute molecules. Their concentration in the vapour phase (the mobile phase) is determined by their vapour pressure but this, in turn, depends on their polarity. In all cases, the polarity of the stationary and the mobile phases must be matched to that of the solute molecules, so that there is sufficient difference between the rates of movement of the components in the sample to achieve separation. So for column chromatography, which is run until components are eluted from the end of the column, the characteristic rate of movement of a particular solute will be shown in its particular retention time ($t_R$) or its elution volume ($V_e$). In planar chromatography, which runs for a definite time, a particular solute will travel a characteristic distance but this is usually referred to in terms of the $R_f$.

In some analyses you will have a series of very similar molecules – hydrocarbons in oils or sugars extracted from beet perhaps. In such cases the order in which they are eluted from the column is largely dictated by their relative molecular mass, so you can expect that larger molecules will have a longer retention time or greater elution volume. In hydrocarbon analysis or with straight-chain alcohols you can even show that the retention time depends on the logarithm of the relative molecular mass. However, this sort of correlation is unreliable if there are isomers in the sample.

We defined these terms in Section 5.1 (see Figures 5.1 and 5.3). Obviously, the values of all of these characteristics will depend on the dimensions of the system and the flow rate that you use. If we are going to get values which can be compared between one laboratory and another we must have measures which are independent of the system. In column chromatography we might use the Kovacs index or the McReynolds number, but we do not need to consider them here. You should refer to more advanced treatments of chromatography for details.

## 5.3 Looking at chromatograms

Remember that isomers are substances with the same formula (and hence the same relative molecular mass) but with different structures.

Although we have already given a diagram of a typical chromatogram (Figure 5.1) it is useful to give a slightly modified version here (Figure 5.5) so that we can describe in more detail what you should look for in the chromatogram and how you can interpret it. In this chromatogram we have shown time (or elution volume) running from left to right, as you might expect. You will find, however, that many chromatograms are presented with the time or volume axis running from right to left. This is because of the way in which many chart recorders are set up. It does not, of course, affect the theory but it might lead you to misinterpret the chromatogram. You must check whether it is left to right or right to left.

If the direction of the chromatogram is not clear from the labelling of the axes you can use these clues to help you decide: if there is a large solvent peak it will be at the beginning; early peaks are narrower, while later peaks are broader because they have more time for diffusion.

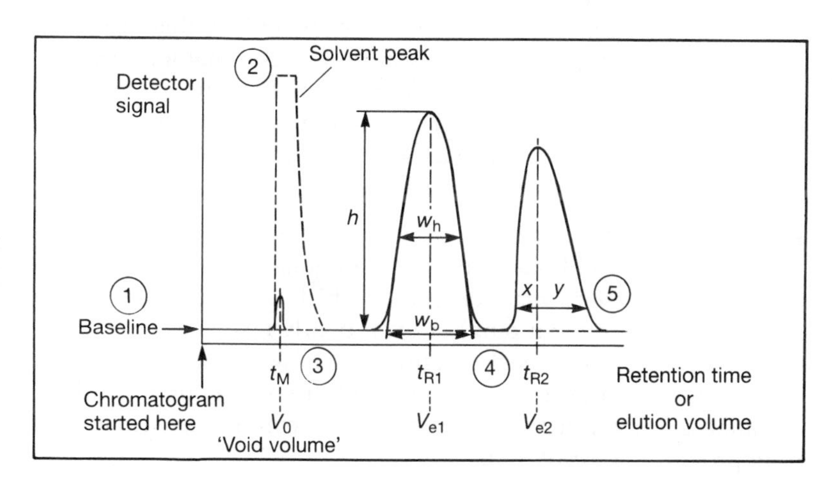


**Figure 5.5** Chromatography parameters.

Here are some comments about Figure 5.5.

1. The baseline is usually set a little above the edge of the chart to allow for drift. If it goes below, the pen just runs along the edge of the chart and you do not know whether any peaks have been missed.
2. Very large peaks will not be fully recorded because the pen is attempting to go off the top of the chart, hence the flat top to the trace. If this happens for a solvent peak it is unlikely to matter, otherwise you must readjust your sample size or the settings on the instrument amplifier so that the whole peak can be seen.
3. A short time after you have started the chromatogram you may get a peak due to non-retained molecules which travel at the same rate as the mobile phase. In gas chromatography this may be due to a trace of air or it may be some solvent in the sample. If it is solvent it will be a large peak (the dotted line in the diagram). With liquid chromatography it will most probably be because the sample solvent is somewhat different from the mobile phase. You can use this peak to measure the time taken ($t_M$), or the volume required (the 'void volume', $V_0$), for the mobile phase to sweep through the column length. No genuine peaks from your sample can appear before this one.

Why does the solvent give a large peak?

4. $t_{R1}$ and $t_{R2}$ are the retention times of the peaks and you hope that each represents a single component of your sample. If the peaks are not narrow and symmetrical you may suspect that there is more than one component in the peak. However, see note 5. (The retention time is measured from the time of injection, but for some purposes it is necessary to use the adjusted retention time ($t'_R$) which is measured from the time $t_M$.)
5. Quite often peaks show more curvature on the descending side. Such a peak is an asymmetric peak and, in this case, it is said to be tailing. It is a sign of imperfect chromatography and can be due to too much affinity between the component and the stationary phase. (The amount of asymmetry can be quoted as the asymmetry factor, $A_s$, which is the ratio $y/x$ and is measured at 10% of the peak height.)

**Resolution** is the term describing how well two peaks are separated. You will hope your chromatogram will show complete resolution but there may be peaks which are incompletely resolved (partially overlapping) or even poorly resolved (a shoulder on one side of a main peak). Figure 5.6 shows these cases. If you cannot get the information you want from a poorly resolved chromatogram you have to try using other conditions with a view to improving the resolution.

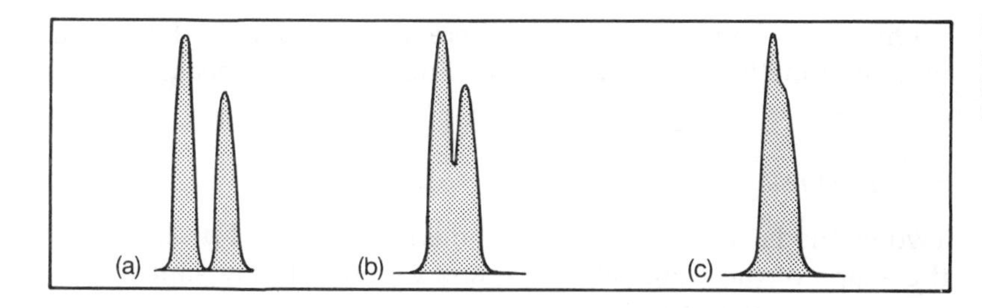


**Figure 5.6** Resolution of peaks in a chromatogram. (a) Good resolution; (b) partial resolution; (c) poor resolution.

### 5.3.1 Methods of peak area measurement

We have already mentioned (Section 5.1) that quantitative information can be obtained from chromatograms. The area between the peak and the baseline is a measure of the amount of a component in the sample.

First you must **establish where the baseline is**. Ideally it will be a flat line near the bottom of the recorder chart, as shown in Figure 5.5. We expect the recorder trace to return to this line in between samples, and also, as far as possible, between the peaks while a sample is running. However, there are several reasons why in practice the baseline is not found to be flat.

- The line may gradually 'drift' up or down, perhaps due to changing conditions during a run.
- Some components of a sample may be tailing severely.
- Some material may still be eluting from previous samples.
- The conditions being used may be too severe for the stationary phase ('column bleed').

Also, if there is incomplete resolution, you have to make a judgement as to the most appropriate division of the overlapping peaks. Figure 5.7 illustrates these possibilities.

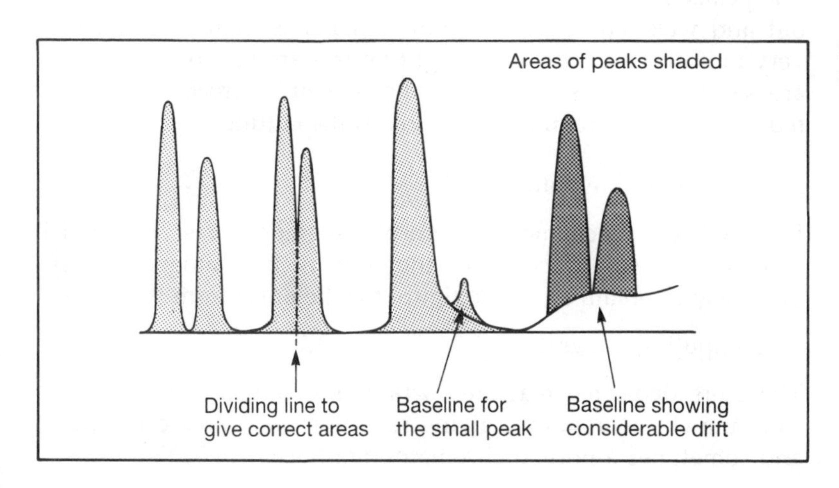


**Figure 5.7** Judging the baseline and areas of chromatogram peaks.

Once you have established where the baseline is there is a variety of methods you can use for finding areas of peaks.

*Rough and ready*

Two methods are based on the geometry of the ideal peak. (It has the shape of the Gaussian curve and an area closely similar to a triangle of similar dimensions.)

1. Half base multiplied by height: $\frac{1}{2}(w_b \times h)$ in Figure 5.5

Note that the base is the segment of the baseline formed by the projection of the straightest portions of the sides down to the baseline. Also, the height is that of the peak rather than the triangle. The error in this is not significant unless the peak is quite rounded. However, you may have difficulty judging the correct position of the sides of the triangle.

2. Height multiplied by width at half height: $(w_h \times h)$ in Figure 5.5

For a true triangle this is geometrically equivalent to the first method. For chromatography peaks it is quicker than the first method but perhaps a little less precise. The main source of error is in the measurement of the width since this may only be 1–2 mm. You can run the chart recorder at a faster rate to give broader peaks on the paper and this may give greater precision in your measurements.

*More precise*

1. 'Cut and weigh'

The peaks of the original trace, or some accurate copy, can be cut out and weighed. Because modern paper is manufactured to a very uniform thickness the weight figures are proportional to the areas of the cut-outs. This method can be quite precise but is very tedious and does require some skill in the cutting.

2. Mechanical integrators

For a while a mechanical system of integration was used which produced an extra trace beside the original chromatogram. It was very easy to obtain a precise figure for the areas from this trace.

3. Computing integrators

These are dedicated machines which deal directly with the data from the chromatograph, giving retention data, area calculations and sometimes a normal chromatogram trace.

*Fully computerized*

All the above methods are now being rapidly superseded by software packages which can do all the data handling and presentation required by the most sophisticated chromatographer. You are most likely to do the data manipulation and even the setting up of the system on-screen. As with all sophisticated systems, do not expect to be able to use or understand all the features at the first go. You must understand how to deal with the raw data from the chromatograph first.

## 5.3.2 Calibration

Once you have estimated, or got the computer to estimate, the areas of the peaks you may still need to calibrate the system for each peak of interest. If you are dealing with a series of very similar molecules (such as fatty acid methyl esters used in the analysis of fats, or glucose chains of various lengths from the hydrolysis of starch) the detector will respond to the same extent for all the components and the peaks areas can be converted directly to relative proportions such as percentages, as shown in Figure 5.8(a).

For a **simple calibration** in terms of concentration the procedure is very similar to that outlined in Section 1.6. You use a series of standards of known concentration and obtain their chromatograms. You measure the peak areas and plot the values against

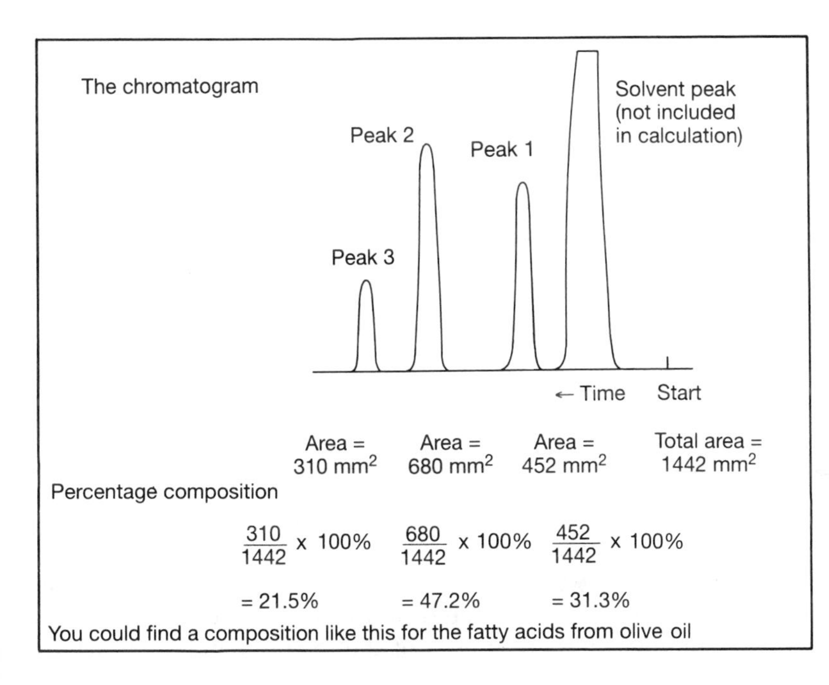


**Figure 5.8(a)** Determining the percentage composition of a mixture of similar substances.

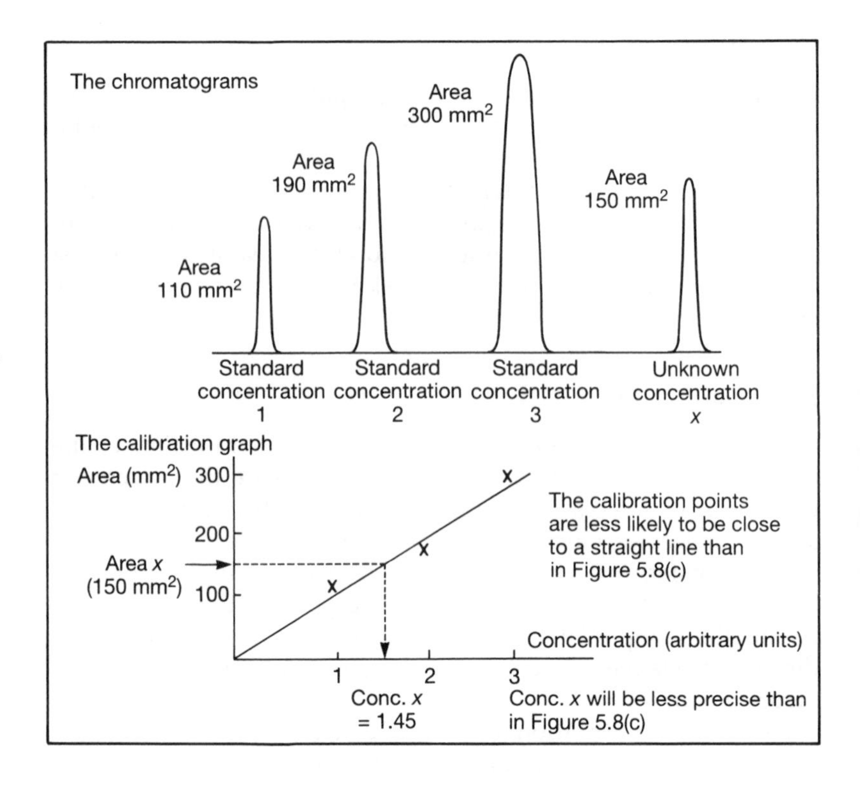


**Figure 5.8(b)** Calibrating peak areas for quantitative chromatography.

the concentrations. You should obtain a straight line for this, or very nearly straight, and you can use this calibration graph to find the concentrations of samples of unknown concentration (see Figure 5.8b).

Simple calibration suffers from the drawback that you cannot be sure the very small volume you inject is precise and so you may get some variation in the actual volume injected. There is a refinement of the simple calibration which uses an **internal standard**. You add an extra component (the internal standard) to your normal standards and samples. The same concentration of this extra component is added to each, giving an extra peak in each chromatogram. The extra peak represents the same concentration throughout although its peak may vary in size if the actual injection volume varies. However, the **ratio** of the peak area of interest to that of the internal standard provides an automatic compensation for variation in the injected volume since both will vary in the same proportion. The values of the ratios are plotted for calibration instead of the areas as such (see Figure 5.8c).

When the detector responds differently to the components in your sample you will need a separate calibration for each peak you are interested in. You could find this situation in an analysis such as finding the percentages of the minor components which

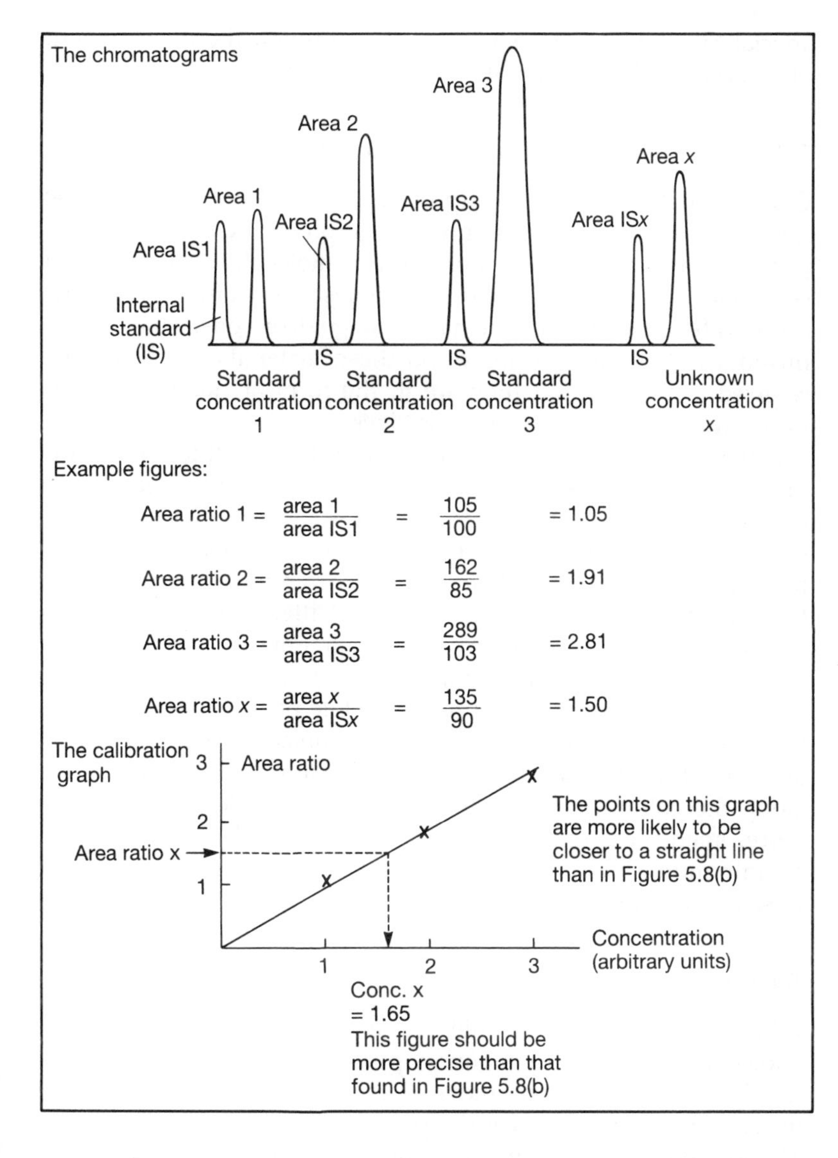


**Figure 5.8(c)** Calibrating peak areas using an internal standard.

affect the flavour of wine samples. It is quite laborious and you are unlikely to do it for more than two or three peaks.

## 5.4 Specialized versions of simple liquid chromatography

For this section, we will consider systems of liquid chromatography which are derived directly from column chromatography, and can still be performed with the simplest apparatus. The specialization has been made possible through the development of stationary-phase materials with particularly suitable structures.

### 5.4.1 Gel permeation chromatography

Gel permeation chromatography does not rely on partition or adsorption and so might be classed separately from other chromatography methods. It is included here because of its superficial resemblance and its similarity in operation to the other methods.

Gel permeation chromatography (GPC) is also known as 'gel filtration' or 'size-exclusion chromatography' (SEC). Its stationary phase is a synthetic polymer with a molecular structure that contains a range of 'molecular-sized holes'. Very often it is a polysaccharide type of material having a gel-like consistency when mixed with the mobile phase. So these materials are frequently referred to as **gels**, but they are in the form of small soft beads rather than a continuous layer or block.

Different-sized solute molecules can diffuse to different extents into the molecular structure of the polymer. Ideally, there is no interaction between the solute and the stationary phase, and the solute molecules also diffuse back into the bulk mobile phase and move along with it, but while they are within the stationary phase they do not move along the column. Different-sized molecules will spend different proportions of their time in and out of the stationary phase. Their average speed of movement along the column is thus less than that of the mobile phase and depends on the size of the molecules. The largest molecules penetrate least and spend most of their time in the mobile phase, hence moving along the column fastest. Similarly, smaller molecules penetrate more, spend less time in the mobile phase and move along the column more slowly. Thus the order of elution is inversely related to size: **largest first, smallest last**.

*Total exclusion and de-salting*

When the size of a molecule is greater than the largest of the 'molecular holes' in the stationary phase the molecule is said to be **totally excluded** and will move at the same speed as the mobile phase. For many biological samples there is likely to be a variety of molecules in this category and they will not be separated from each other. Similarly, molecules whose size is less than the smallest of the holes will diffuse into the gel to the maximum extent and elute together at the end of the chromatogram.

This gives a simple method of removing small molecules and ions from samples in which the main interest is in the large molecules. You may have used ammonium sulphate to isolate a protein sample (the 'salting out' process) or phenol to extract nucleic acids. You could pass the crude sample down a GPC column and retrieve the eluate. The small molecules and ions should be retained enough for the larger molecules to be eluted without them, i.e. **de-salted**.

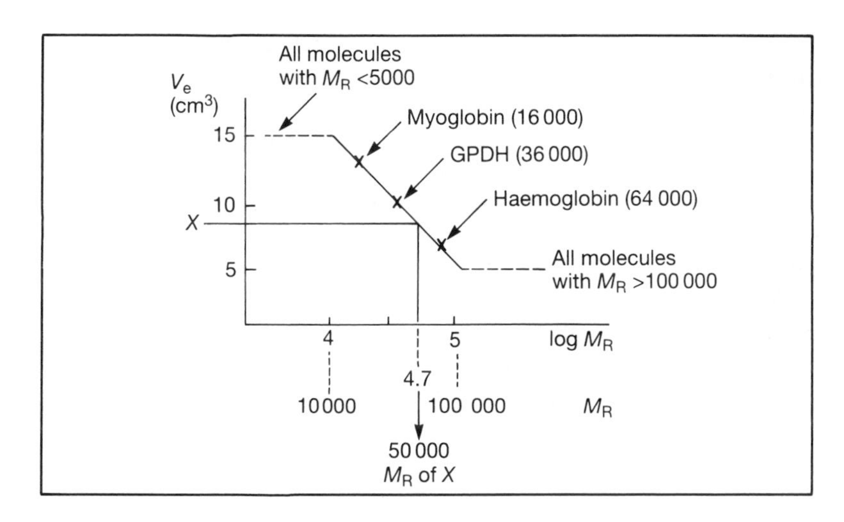


**Figure 5.9** Determining a relative molecular mass by gel permeation chromatography.

*Separation of large molecules*

We can be more precise about the order of elution of molecules that do penetrate the gel but, at this stage, we should point out that we have not yet been precise enough about what we mean by size. We need to consider shape as well. For instance, long thin molecules such as amylose (found in starch) or collagen (in connective tissue) will have different penetrating ability from more spherical (globular) molecules such as glycogen, albumin proteins and many enzymes, even if they are similar in total volume or mass. Provided that we restrict our comparisons to molecules of similar type and shape, we find that the elution volumes are inversely related to the logarithm of the relative molecular mass.

So you can determine the relative molecular mass of large molecules by calibrating the column with samples of known relative molecular mass. Figure 5.9 shows this for some globular proteins. If you had a newly isolated globular protein (X) which eluted at 9 $cm^3$ you could determine that it has $\log M_R$ of 4.7, so its actual $M_R$ is approximately 50 000.

GPC can be used on almost any scale up to about 1 g, but then large quantities of gel are need to get the requisite resolution. Another drawback is that it can be very slow: it may take more than a day to achieve some separations.

## 5.4.2 Fast protein liquid chromatography

To overcome the slowness of GPC other modified polysaccharide-type materials have been developed that are 'macroporous' (e.g. the Sepharose range). They have a much greater surface area of gel in contact with the mobile phase, which allows much more diffusion into and out of the gel. Additionally, the gel beads are of uniform size and a pump can give a much faster flow rate without losing resolution. These stationary phases do not separate on size alone since they are produced in forms (polyethers) which utilize hydrophilic (polar) interactions and in forms (phenyl Sepharose, alkyl Sepharose) which work through hydrophobic (non-polar) interactions.

The name Sepharose is a registered trade-mark of Pharmacia Biotech AB.

These techniques need specially designed column fittings and pumps. They have all parts that are in contact with the sample made of materials that do not denature the molecules in any way. This most usually means plastics (especially PTFE) rather than metals, so only moderate pumping pressures can be used. Nevertheless, separations can be achieved in minutes rather than hours.

### 5.4.3 Ion-exchange chromatography

For this technique the stationary phase has an ionic character achieved by modifying polysaccharide type materials so that they contain the acidic or basic groups which form ions when in contact with mobile phase. The groups generally used are sulphonic acid (which produces $-SO_3^-$), carboxylic acid ($-COO^-$), amine ($-NH_3^+$) and quaternary ammonium ($-N(CH_3)_3^+$). They attract charges of opposite sign on the molecules in the sample. These molecules are retained by the column, while others are rapidly eluted. In order to elute the molecules of interest, the mobile phase is changed so that either the pH is different, or the salt concentration or both. Under the changed conditions the attractions between the sample molecules and the column are weakened, so the molecules of interest are eluted. If you make a gradual change in the elution buffer you can elute the components sequentially, producing a chromatographic effect (Figure 5.10). The long-established technique of amino acid analysis of proteins works on this principle. You hydrolyse a protein, which produces a mixture of 20 or so amino acids; there are then separated by ion-exchange chromatography. In fact, there are now systems which will do the whole process automatically.

### 5.4.4 Affinity chromatography

The 'affinity' in this technique is the very specific attraction between particular combinations of biological molecules such as enzyme–substrate or antigen–antibody.

First the stationary phase is treated with a reagent so that it can further combine with a particular substrate or antibody. As long as the attachment does not destroy the binding capabilities the column will retain the corresponding enzyme or antigen from a

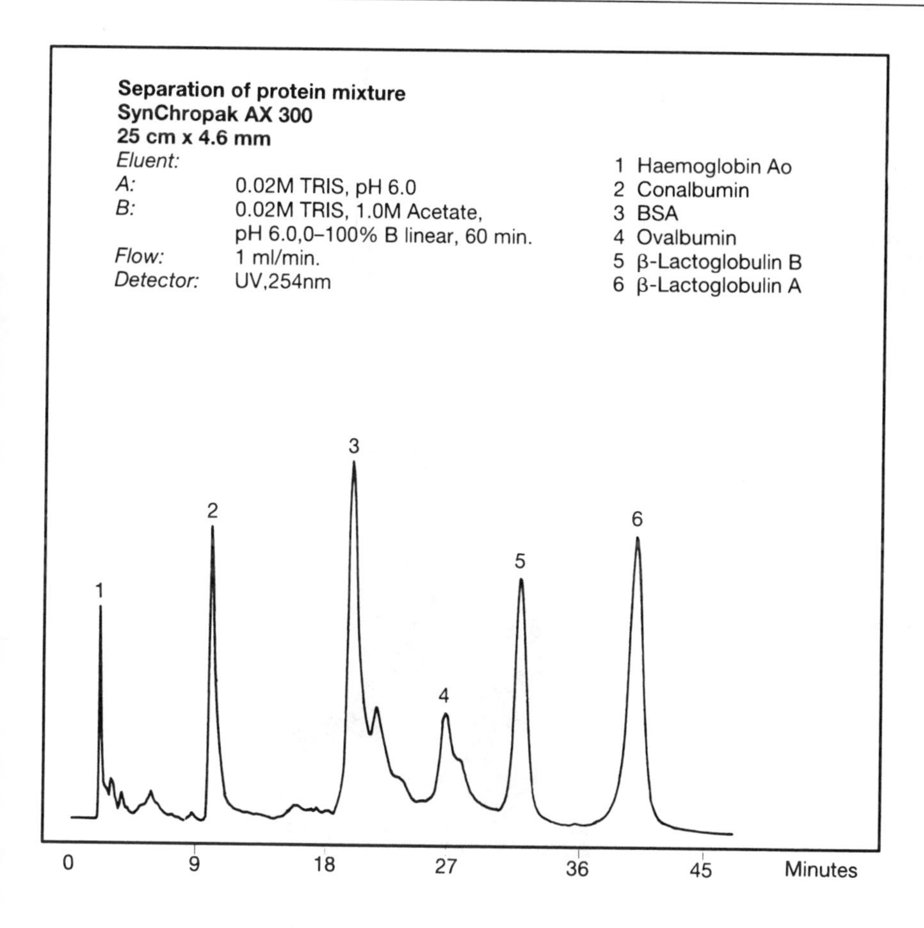


**Figure 5.10** Separation of a protein mixture using ion-exchange HPLC. (Courtesy Fisons, UK.)

sample, while all else is effectively washed through. The substance bound to the column now has to be released. This can be done by reducing the binding to the column by changing the pH or the salt concentration of the mobile phase, as in ion-exchange. Alternatively you can pass a comparatively concentrated solution of the real substrate or antibody, which displaces the retained molecules (Figure 5.11).

## 5.5 High-performance liquid chromatography

The aim of 'high performance' is to produce a very efficient column. That is, one which can separate many components from a complex mixture and resolve components that have very similar retention times or elution volumes. Chromatography theory shows that better efficiency is obtained if the particle size of the stationary phase is made as small as possible and as uniform as possible. A high pressure is required to obtain a useful flow rate, but not excessively so.

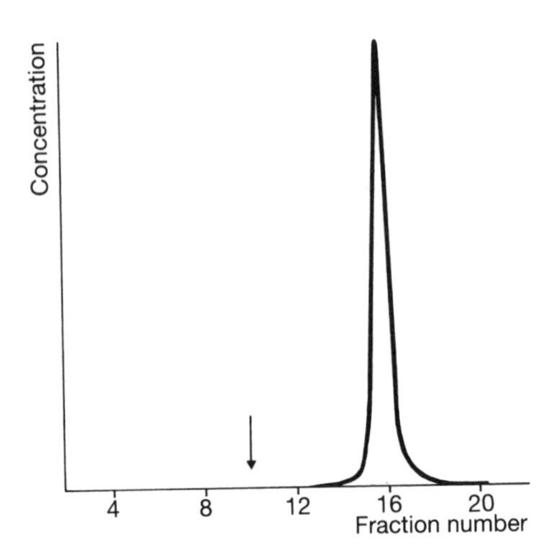


**Figure 5.11** Affinity chromatography of concanavalin A. Purification of concanavalin A on glucosamine immobilized on activated CH Sepharose® 4B. The sample was applied in acetate buffer (0.01 M, pH 5.5, 1 mM $CaCl_2$, 1 mM $MnCl_2$, 1 M NaCl) and the bound concanavalin A eluted by adding glucose (0.1 M) to the buffer at the point indicated by the arrow. (Courtesy Pharmacia Biotech AB.)

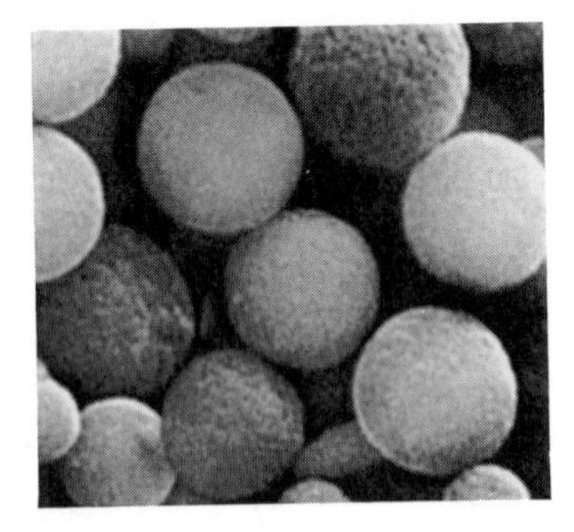

**Figure 5.12** Typical HPLC stationary-phase particles. The particles are approximately 5 $\mu$m in diameter. They show a narrow range of size (monodisperse) and are almost perfect spheres for maximum chromatographic efficiency. (Courtesy Alltech, UK.)

### 5.5.1 Common stationary phases and solvents

*Stationary phases*

Almost all high-performance liquid chromatography (HPLC) uses stationary phases based on silica ($SiO_2$). This is rigid enough to withstand the high pressures involved and it can be produced in the form of microspheres with a variation of size of less than 1 $\mu$m and diameters in the range 3–10 $\mu$m (see Figure 5.12).

Its chemical properties are such that it can be used directly as an adsorbent in normal-phase chromatography, or its surface can be chemically modified to widen the scope of its use and enable us to use reverse-phase chromatography. Commonly the hydrocarbon chains $C_8H_{17}-$ (C8) and $C_{18}H_{37}-$ (C18 or ODS, which stands for octadecylsilyl) are linked through a silyl group to the silica. Other chains similarly linked include aminopropyl and cyanopropyl.

The silyl link to the silica is fairly stable in mildly acidic or neutral conditions but is readily hydrolysed when the pH is below about 2 or exceeds about 8, so you have to make sure that the mobile phase is within this range. Even so, there is always some physical or mechanical degradation of the stationary phase whatever conditions you use. It sometimes leads to a shrinking of the stationary phase, producing gaps at the ends of the column ('voids') which reduce the effectiveness of the column.

*Solvents*

The most widely used system is reverse phase (see Section 5.2.2) and so polar solvents are used, such as water or methanol, or moderately polar ones such as acetonitrile and tetrahydrofuran (THF). You often use mixtures and if the composition is constant the elution of the chromatogram is known as **isocratic**. However, when there is a wide range of polarities in the components of sample you may use a **gradient**, i.e. a varying composition of mobile phase controlled by a programmed valve connected to the solvent reservoirs. Gradients in reverse-phase systems start with higher polarity and become less polar as the elution proceeds. Sometimes a pH buffer is included in the mobile phase to ensure that the sample components are either completely ionized or completely unionized, whichever is best suited to the column. You must make sure that the buffer is soluble enough in the mobile phase that no crystallization occurs, otherwise the system can become blocked or the pumps damaged.

Since HPLC is a highly sensitive technique, the solvents must be particularly well purified. Manufacturers now usually provide a special 'HPLC-grade' of solvents, even of water! These solvents are as free as possible from UV-absorbing materials since this is the most widely used method of detection. Also, solvents must be rigorously filtered (e.g. down to 0.2 $\mu$m) so that there is no particulate matter to block or damage the system.

Before the mobile phase enters the system it is degassed. If this is not done, bubbles can appear in some parts of the system because of the high pressures and the pumping action. Bubbles can spoil the flow and produce fluctuations in the pressure which may affect the detector. There are several methods of degassing the mobile phase, including ultrasonic treatment, vacuum treatment and bubbling helium through it ('sparging').

### 5.5.2 A typical HPLC set-up

Almost all systems are modular and many of the units are boxes of about 30 × 20 × 10 cm. Nevertheless, individual set-ups may look rather different, with the units and their configuration chosen to suit the user.

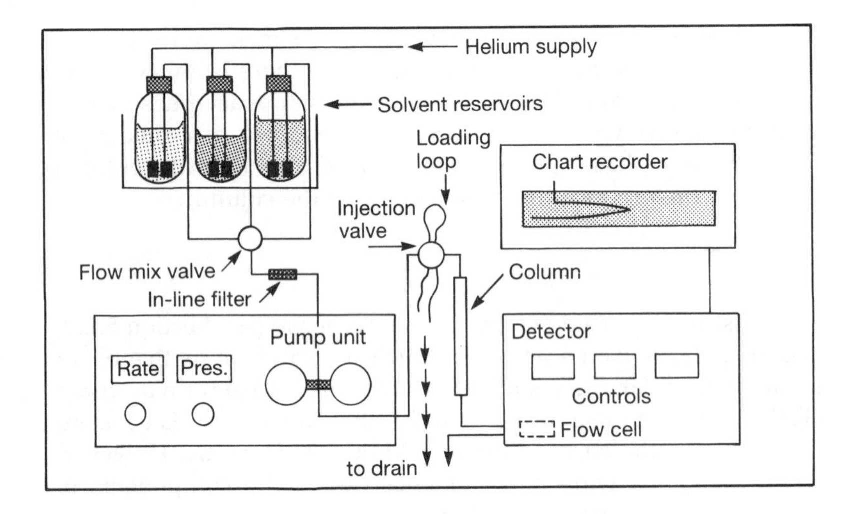


**Figure 5.13** Typical HPLC set-up.

**The solvent reservoirs** may contain ready-mixed or several pure solvents. They may be degassed by bubblers attached to a helium supply. For gradient elution the composition of the mobile phase entering the pump is controlled by a computerized switching valve which enables different proportions of the solvents to be admitted to the solvent line. An in-line filter is included as a precaution.

**The pump** often has two pistons with 'pulse dampening' to give a uniform flow. The flow rate is typically in the range $0.5–2\,cm^3\,min^{-1}$ ('ml per min'). A pressure monitor reduces the flow rate if the maximum is exceeded. There may be a purge facility which by-passes the column to allow a rapid change of mobile phase in the tubes before the column.

**The tubing and joints** are made to withstand the high pressures; they have an internal diameter of less than 0.5 mm. They are usually stainless steel so you must be aware that it might denature the molecules in your sample. Also, it can be corroded by chloride ions. There are plastic alternatives but these are subject to pressure limitations. You must, of course, check for leaks.

**The injection valve** is a heavy-duty two-position valve. It is operated so that you first place a portion of sample into an isolated loop (**load** the loop) while the mobile phase is directed straight onto the column, and then turn the valve to redirect the mobile phase through the loop so that you **inject** the sample onto the column. There is no interruption in the flow or the pressure in this operation.

**A guard column** may be fitted before the main column. It is a small column containing the same stationary phase as the main column but of larger particle size. It removes strongly retained components from the sample and so prevents contamination of the main column.

**The column** housing has a highly machined mirror-finish interior to give the best flow of the mobile phase. Typical dimensions of the column are 250 × 4 mm and 100 × 4 mm.

**The detector** must be highly sensitive and stable. The flow cell volume must be compatible with the resolution required, otherwise remixing of the separated components may occur. Its volume is typically about 8 $mm^3$ ($\mu l$).

### 5.5.3 Types of detector

Detectors may be broadly classified as those which detect almost any component (universal) and those which use more specialized forms of detection (selective). The latter are used when sufficient sensitivity is not obtainable by a universal detector or when some selectivity is desired, such as when only some of the components of the sample are of interest.

*Universal*

**UV absorption** detectors work just like a single-beam UV spectrophotometer (see Section 3.3.1). They are the most widely used detector since many biological compounds absorb UV radiation, particularly at short wavelengths (190–220 nm). You may find, however, that it is difficult to calibrate the system and maintain a stable baseline at these wavelengths, and solvent absorption may interfere. The most common detectors allow you to select only one wavelength for an analysis but you are likely to have a variety of $\lambda_{max}$ values for the components of the sample and your choice of wavelength may have to be a compromise. The dual-wavelength system overcomes this to some extent by rapidly switching between two $\lambda_{max}$ values. The most advanced system is the diode array, which detects all the wavelengths simultaneously (see Section 3.3.5), so you can get spectral information for each component. This helps in the identification of components.

**Refractive index** detection registers the changes in refractive index as the components of the sample are eluted. It is useful when components have little or no UV absorption even at short wavelengths. However, it is very sensitive to changes in temperature and to changes in mobile-phase composition. Because of this, it cannot be used for gradient elution.

**Mass detection**: a recent innovation evaporates the eluate in a special chamber. If a component is present it produces a residual 'smoke' which is detected and quantified by a light-scattering system. It is claimed to be a truly universal detector.

**Box 5.4 Practical tips for HPLC**

**Do:**

- check that you have an adequate supply of degassed solvents;
- check that mixtures of solvents are thoroughly mixed;
- look out for leaks and erratic pressures;
- check that the sample is soluble enough in mobile phase;
- filter your sample down to 0.2 μm before injecting it;
- run standards separately and as mixtures before test samples;
- before leaving a set-up, ensure that the column has been thoroughly flushed to remove buffers and any residual solutes.

**Don't:**

- put crude samples through without a guard column;
- let solvent reservoirs go dry;
- make arbitrary or drastic changes to the running conditions (always have a good reason for what you are changing and some idea of possible outcomes);
- change the flow rate rapidly;
- leave the column without flushing.

**Trouble-shooting**

You often find things go wrong with chromatography systems and there are all sorts of reasons why. You should have some idea of what results you expect then, if they are not as you expect, ask yourself whether the trouble might be due to the sample, the column, the solvent or the instruments. Be logical in your questions and deductions.

Here are a few troubles and their remedies; manufacturers' information and specialized texts give much fuller lists, and computer programs are now available to help work your way through the many possibilities.

| *Symptom* | *Possible diagnosis* | *Remedy* |
|---|---|---|
| No peaks | No sample<br>Detector off<br>Detector not connected<br>Wrong injection sequence | |
| Wide peak in several narrow ones | Late-eluting component from previous injection | Give more time or change solvent |
| Unusual shapes of peaks | Wrong type of column<br>Voids in degraded column | Change column |

*Specialized detectors*

**The conductivity** of the mobile phase varies with the type and concentration of ions, so it is particularly useful for ionic samples and in ion chromatography for the separation of, say, chloride, bromide, iodide and sulphate.

**Electrochemical** detection is based on the methods described in Chapter 4 and is for substances that can be oxidized or reduced such as sugars or vitamin C. The electrodes are fitted into a low-volume flow cell.

**Fluorescence** (described in Chapter 3) is very sensitive for some compounds, quinine for example, and is quite selective but subject to interferences (quenching).

**Optical activity** is the rotation of the plane of polarized light produced by asymmetric (chiral) molecules. It enables you to detect which optical isomers you have and in what proportions, which is important in drug studies.

**Radioactivity** detectors incorporate a solid scintillator in the flow cell and detect the light emission by a photomultiplier tube. They are used when it is important to know the amounts of radioactivity in the various components, as in metabolic studies.

### 5.5.4 Applications

Almost anything can be analysed by HPLC, although, of course, it is not always the most appropriate method. However, there is such a wide range of applications, with adequate or good separations and sensitivities, that it is one of the most widely used analytical methods. Here are a few examples:

- the levels of caffeine in tea and coffee and other drinks;
- sugars in honey or fruit juices: which ones are present and in what proportions;
- the artificial sweeteners and preservatives in soft drinks can be checked to ensure that they conform with national legislation;
- vitamin levels in foodstuffs can be determined more rapidly than by microbiological tests;
- drugs of various sorts: their identification for forensic purposes or whether they have been used in the doping of race horses;
- steroids: detection in athletes' body fluids to prove that performance has been artificially enhanced;
- amino acid levels are abnormally high in the blood in some disease conditions; detecting this can help in diagnosis.

### 5.5.5 High-performance thin-layer chromatography (HPTLC)

The materials developed for HPLC can also be used in TLC and give the same benefits of greater efficiency of separation and the wide applicability of the reverse-phase technique. However, it is necessary to apply the sample spot in as small an area as possible so as not to degrade the advantages of the method. This has to be done mechanically to achieve the necessary precision. With the increased resolution of the technique, the traditional 20 × 20 cm plate can be replaced by a smaller size and so development times are also shorter than the 1–3 h of normal TLC. They become similar to the times needed for HPLC and there is an advantage that several samples can be run simultaneously.

Detection is done similarly to normal TLC but to maintain the precision and enable quantitative measurements to be made it is normal to use a scanning densitometer. This, of course, increases the cost considerably over normal TLC, but when compared with HPLC some workers consider it superior in resolution, ease and speed of use.

## 5.6 Gas chromatography

We have left discussion of gas chromatography (GC) until last since its differences from simple liquid chromatography are rather greater than the methods so far described. However, it has been a routine technique for longer than HPLC or HPTLC and so may often be regarded as being more familiar to analysts. In its early days it was sometimes known as 'vapour phase chromatography' (VPC) but this name was superseded by 'gas–liquid chromatography' (GLC) since most stationary phases were non-volatile liquids on support materials. Today there are completely solid stationary phases (made of porous polymers) and so there is also 'gas–solid chromatography' (GSC). However, it is now usual to refer to any sort of chromatography involving a gaseous mobile phase as 'gas chromatography'.

### 5.6.1 The gas chromatograph

The basic equipment is shown in Figure 5.14 and has components with similar functions to those of HPLC. The look of the system may vary from one manufacturer to another but it is generally housed in a roughly cubical box. The chromatogram itself has traditionally been drawn out using a chart recorder but more recently there has been the addition of a computer to produce the print-out and to manipulate the data.

**Carrier gas supply**: the gas must be inert under the conditions used. It is important to have traps to remove traces of oxygen and

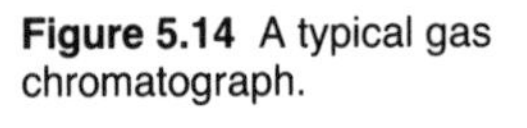
**Figure 5.14** A typical gas chromatograph.

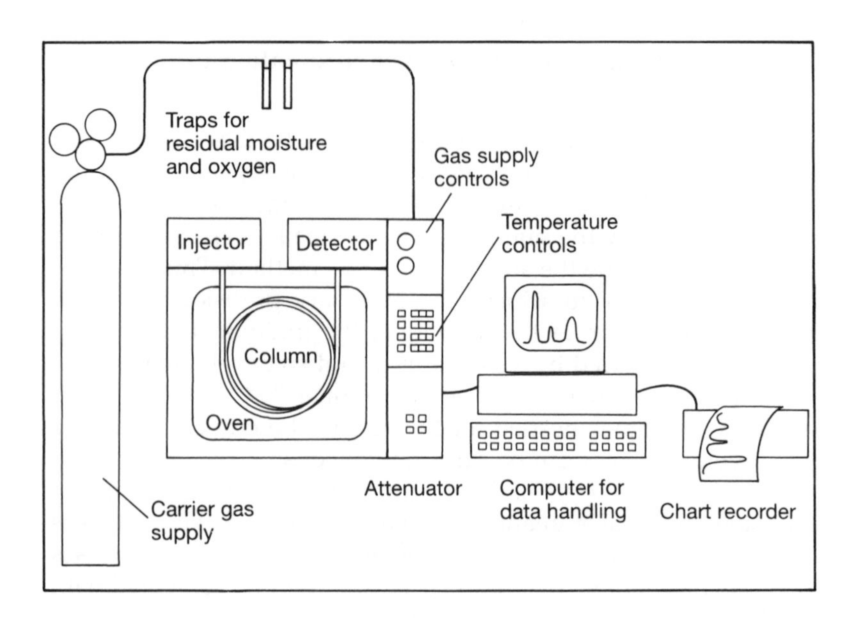

water. The most common gases are nitrogen (for packed columns) and helium (for capillary columns).

**Pressure regulator**: since gases are compressible the pressure does not give a direct measure of the carrier gas flow rate, but a routine operator will soon get to know the values which give appropriate flow rates for a particular machine.

**The flow rate** is generally monitored at the end of the column, before starting up the whole system. Often a bubble type of flow meter is used but there are other models.

**The injection port** is heated to a temperature about 10°C higher than that of the column to avoid condensation. If you are using a routine packed column your sample will be a dilute solution in a volatile solvent. You make the injection using a modified hypodermic syringe with a long needle to reach down to the level of the stationary phase in the column. Only a very small volume is injected, typically about $1\,mm^3$ ($1\,\mu l$), and it vaporizes at the moment of injection. The seal is maintained by a silicone rubber septum which is punctured with each injection but reseals on removal of the syringe needle because it is under compression. However, it has a limited life and must be checked for leaks. There are special injectors for gas samples and for capillary columns.

**The oven** is fan-assisted to give very even and precisely controlled temperatures. It has a low heat capacity to allow rapid raising and lowering of temperature.

**The detector** is, like the injector, heated to a temperature higher than that of the column to avoid condensation. Various types of detector are described below. It should give a linear response to the concentrations of the components but it can be overloaded and this may show up as the tops of the peaks being flatter than normal.

**The attenuator** adjusts the strength of the signal to the chart recorder or VDU so that complete peaks may be recorded. The amount of attenuation varies with the sample size and the sensitivity of the detector response. Higher settings of the attenuator are needed for larger samples.

### 5.6.2 GC columns and stationary phases

There are two sorts of column used in gas chromatography. At first you are more like to use the type which contains the stationary-phase powder packed inside. Later, you may use the capillary type, in which the stationary phase is a coating on the inside of the column.

**Packed columns** are 1–2 m long in the form of a coil 10–20 cm diameter and the width of the column is about 1 cm. It is made of glass or metal.

**Table 5.3** Some typical GC stationary phases

| *Name* | *Character* | *Typical applications* |
|---|---|---|
| Squalene<br>Apiezon | Non-polar | Hydrocarbons, oils, petrols |
| Diethyleneglycol succinate (DEGS) | Moderately polar | Fatty acid methyl esters |
| OV1 (methyl silicone) | | Fruit flavours and other essences |
| Polyethylene glycol (carbowax)<br>Poropak Q (solid) | Polar | Alcohols, low molecular weight acids |

Glass columns are more inert towards the sample and they allow you to check the contents. The column packing should be intact since gaps lower the efficiency of the column. Most column packings are white and if there is deterioration of the stationary phase you may see some discoloration. Remember, however, that glass is fragile and needs particular care in installation and removal from the chromatograph.

Metal columns are more robust, but they may catalyse decomposition of the sample while the chromatography is taking place. The most commonly used metals are stainless steel, which may react with chlorine-containing compounds such as pesticides, and copper, which may act as a redox catalyst. You cannot inspect the contents of a metal column.

The **stationary phase** is often a non-volatile liquid or low-melting solid and has to be coated onto an inert support material; it may be 5–10% of the combined weight. The support material should not participate in the chromatograph process in any way and the stationary phase should not produce any vapour. Of course, no material is ideal and so it is important to use temperatures within the working range stated by the manufacturer. Typical materials are listed in Table 5.3. Many have complex structures whose precise composition may be different from one manufacturer to another. They are often known by their trade-names and, as with most chromatography, it is difficult to get exact reproduction of the results obtained on other columns even if they are nominally the same.

The benefit of **capillary columns** is in the greater resolution obtained. Length for length their performance is about the same as for packed columns but they are much longer. They may be 10–30 m long (occasionally even 100 m), coiled, with a diameter of about 20 cm and supported on a wire rack. They are made from fused silica and have an internal diameter of 0.2–0.5 mm. The materials used for the stationary phase are the same as for a packed column but are applied to the wall of the column because a packed column of this sort would require too high a gas pressure to give a suitable flow rate. The sample size for such 'wall-coated open tubular' (WCOT) columns has to be extremely small and an alternative has a very fine layer of support material on the wall (SCOT, 'support-coated open tubular') which gives a higher capacity. In either case special injection systems are needed.

### 5.6.3 Detectors

A wide variety of means of detection is used but two are much more common than the rest.

The **flame ionization detector (FID)** has a small hydrogen flame burning at the end of the column. When carbon compounds leave the column and enter the flame ions are produced and a minute current flows between two electrodes connected to the flame. This is amplified and sent to the chart recorder or a computer (or both). This type of detector is almost universal in its applicability since almost all the compounds analysed by GC are organic and hence contain carbon. It can detect a very wide range of levels of concentration. A gas chromatograph using this detector has to have supplies of hydrogen and air as well as the carrier gas, which makes setting up the system a little more complicated.

**Electron capture (ECD)** is a selective detector since it responds mainly to compounds containing chlorine, although compounds containing bromine or oxygen are detected to some extent. It uses a small radioactive source (e.g. nickel-63) to produce electrons which give rise to a current in a detector electrode. When chlorine compounds pass through the detector they capture the electrons, which reduces the current and so a signal is registered. This type of detector is particularly useful for chlorine pesticides since they can be detected without having to extract them from other organic materials which may be in the sample.

Other types of detector have particular applications and details can be found in more advanced textbooks on instrumental analysis. They include thermal conductivity detection (TCD), flame photometric and mass spectrometric detection.

### 5.6.4 Using a gas chromatograph

You will need help to start with but you should know the outline procedure.

1. Select and fit a suitable column.
2. Switch on the carrier gas supply and check the flow rate.
3. Switch on the power and adjust the temperature settings for the column, injector and detector.
4. If you are using FID, switch on hydrogen and air and then light the flame.

5. Switch on the chart recorder and set its zero.
6. When the temperatures are stable, inject a trial sample and adjust the attenuator to produce a satisfactory peak size.
7. Inject other samples and make adjustments to the settings as necessary.

**Box 5.5 Practical tips for GC**

**Do:**

- check your gas supplies and operating conditions for stability;
- check FID flame;
- run some standards first;
- ensure samples are properly processed and are free from non-volatile substances, acid vapours and other contaminants that will adversely affect the column;
- practise your injection technique.

**Don't:**

- heat the column without the carrier gas flowing;
- use conditions outside the recommended range for the stationary phase;
- forget the first peak is probably solvent and is likely to be much larger than the others;
- make a further injection until you are sure the previous sample is completely eluted;
- switch off the carrier gas flow until the column is cool.

**Trouble-shooting**

A few suggestions only; for more detail consult specialized texts and manufacturers' literature.

| *Symptom* | *Possible diagnosis* | *Remedy* |
|---|---|---|
| No peaks | No gas flow (cylinder empty or not turned on?) | |
| | Detector not working | FID: relight flame? |
| | Detector not properly connected to recorder | |
| Wrong number of peaks | Lack of separation | Change conditions or column |
| | Peaks from previous sample still eluting | Give longer time from each run |
| Drifting baseline | Temperature not constant | Check oven settings |
| | Gas flow not constant | |
| Varying retention times | Leaks | Change septum<br>Tighten joints |

### 5.6.5 GC applications

Fatty acids are widespread, especially in the triglyceride lipids. Many analyses have been done by converting the triglycerides to fatty acid methyl esters (FAMES) and then using GC. You can easily determine which fatty acids are present in the lipid sample and in what proportions. You can thus check the amount of unsaturation in edible fats and determine its suitability in certain diets.

Some low molecular weight acids, such as malic acid, are found in fruit juices. Special columns are available for these. However, acids which decompose on heating such as citric acid are not suitable for GC, whatever the column.

Flavours and essences are complex mixtures. They may have a few major components but there is likely to be many minor components and it is these that make it so difficult to mimic flavours artificially. Capillary GC will show the presence of these numerous components but it is virtually impossible to identify them all.

The percentage of ethanol in alcoholic drinks can be determined and a similar system can be used to check the results of a breath-alyser test.

Pesticide residues in plants, animals, water systems and soil can be extracted and then analysed to find whether the levels are permitted or not.

The oil and petrochemical industries make great use of GC. Although these compounds are not biological as such, they are ever present in today's environment and their analysis is as important as any.

## 5.7 Choosing the most appropriate system

### 5.7.1 You have got to start somewhere!

With such a wide variety of interdependent factors it is undoubtedly confusing at first to know how to pick the correct system for a new sample. In many cases you will, of course, be guided by what has been done before with similar samples. This may be in the form of well-established methods developed in a particular laboratory or methods described in research literature. In all cases, however, you can never be quite sure that you can repeat what has been reported. Your system may not perform exactly as described or your sample may be just 'that little bit different'. Thus it is worth considering how the choice of system can be rationalized. The following approach should simplify the procedure and help to avoid unsuitable choices.

### 5.7.2 Gas or liquid chromatography?

One simple criterion for this choice is the volatility of your sample. If the sample can be vaporized without decomposition at temperatures up to about 350°C then GC is a possibility. This choice is limited to molecules whose relative molecular mass is less than about 1000 and whose polarity is comparatively low: fatty acid methyl esters, chlorine-containing pesticides, alcohols and steroids, for example. If you have a particularly small amount of sample available you should also consider whether you need to do further analysis on it since the most widely used detector for GC, flame ionization, is destructive.

For particularly complex mixtures (e.g. food aromas, oil residues) a capillary column system may be needed to achieve the required resolution. However, this is a more advanced technique than using a packed column and you probably would not use it for preliminary work. Polar substances such as sugars and amino acids, however, do not produce a sufficient vapour pressure for GC and they decompose at the temperatures required (150°C or more). Nevertheless, you can use GC for them if you reduce the polarity and increase their volatility by chemically converting these substances to non-polar derivatives ('derivatization'; see below). High relative molecular mass compounds (proteins, polysaccharides and nucleic acids) cannot be volatilized, whether derivatization is used or not, so liquid chromatography must be used (unless you turn to electrophoresis; see Chapter 6).

Just because a sample is a candidate for GC does not mean that you cannot use liquid chromatography. It is now used for all sorts of samples. If you do choose liquid chromatography the type of liquid system requires further choice. Ordinary open-column chromatography as described in Section 5.1 is really only useful for crude separations of relatively large samples. However, you may well do some preliminary investigation of your samples by ordinary TLC since it is simple, cheap and fairly quick (especially if done on a small scale such as on microscope slides). For more refined work, the choice will be between HPLC and HPTLC. Both require expensive instruments and so your choice will be influenced by what is already available in the laboratory. Both are very widely applicable and should give useful results for almost any sample.

### 5.7.3 Derivatization

Each technique can be made more widely applicable by reacting the sample in some way and doing the chromatography with the chemical derivatives. As long as the reaction is done by a standardized procedure and gives reproducible results, the derivative chromatogram can be interpreted in the same way as the original sample.

In GC the derivatization is invariably done before the injection of the sample and is for the purpose of increasing the volatility of

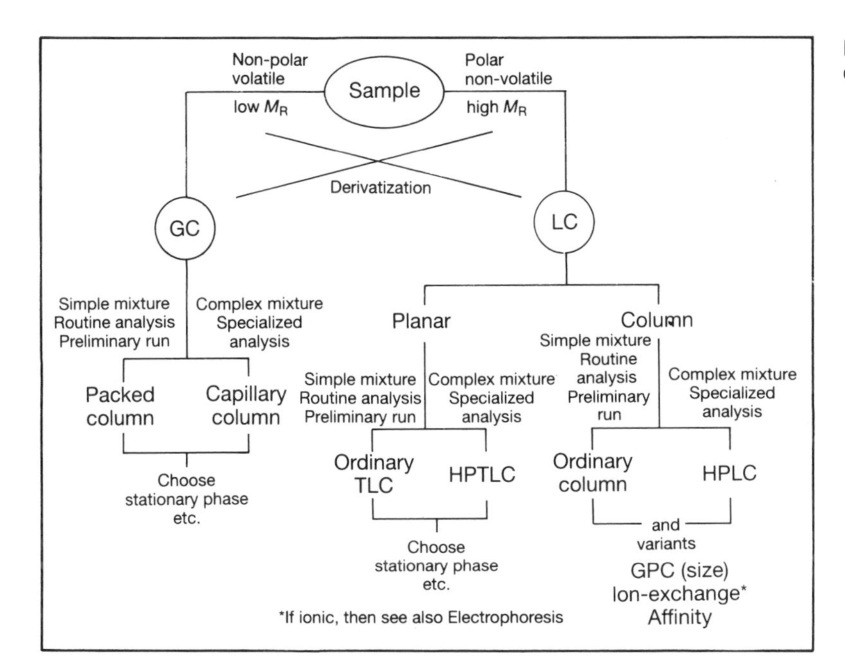


**Figure 5.15** Choices in chromatography.

the sample. Commonly used procedures include **silylation** and **trifluoroacetylation**. Reagents for doing these conversions rapidly under mild conditions are available and so there is little extra work involved. Trimethylsilyl derivatives of amino acids and sugars are sufficiently volatile and stable to be analysed in this way.

With liquid chromatography, the derivitization may be 'pre-column' (before the injection) or 'post-column' (after the components have been eluted). It has generally been used when a suitable detection method was not available. In particular, although UV detection is useful for many samples, sugars and most amino acids are detected only with difficulty. However, by introducing a chromophore (a chemical group that increases the absorption properties of molecules) into the structure, these molecules are detected more readily.

This type of procedure is parallel to that used in colorimetry (Section 3.1.3) when colourless substances are converted to coloured ones so that they can be analysed.

If you use pre-column derivatization you may have to reconsider which column or which solvent you should use. If the derivatization is post-column the necessary mixing of the reagent with the eluate (as well as any heating and allowing time for the reaction) may lead to a loss of resolution (spoiling of the separation obtained). Post-column derivatization units have to be specially designed to avoid this.

### 5.7.4 Choosing the column and the stationary phase

Whichever system you decide upon, gas or liquid, there will be further choices to be made. Most of these will be specific to the system but the following general principles will apply to most chromatography.

The stationary phase and the mobile phase must have a polarity suitable for your sample. You can consult manufacturers' catalogues which usually contain a lot of information about such choices, or you can refer to scientific papers in journals for specialized applications. If there is any doubt, then start with a column of moderate polarity and wide applicability. Such a column for GC could have a methylsilicone stationary phase with nitrogen as the carrier gas (the mobile phase). In reverse-phase liquid chromatography the most commonly used stationary phase is octadecylsilyl silica (ODS) and the mobile phase may well be a methanol/water mixture.

If you cannot get sufficient separation with general-purpose columns in chromatography you will have to consider whether your sample has particular characteristics of polarity, molecular size or type of structure that suggest that you should use the more specialized systems. Whatever system you do use, you must expect to vary the conditions somewhat in order to get the best results you can.

## 5.8 Getting the best from a chromatography system

Getting the best performance can be as much an art as a science! As you would expect, accumulated experience is a great help. Nevertheless, you can always be more efficient if you apply some scientific logic. The process is often called **optimization**.

For a start you must have a clear idea of what factors you can usefully vary. These will include:

- temperature
- flow rate
- composition of the mobile phase
- nature of the column.

It is best to change the easiest things first and go on to more involved changes only if simple changes fail to give acceptable results. An experienced chromatographer will be able to make inspired guesses and take short cuts but the following order is better if experience or information is lacking.

### 5.8.1 Temperature and flow rate

For GC the temperature is easiest to change and it can be done precisely. If peaks emerge so rapidly that they cannot be distinguished a little lowering of the temperature is required. If they take so long that they are very broad you can increase the temperature, as long as you do not exceed the maximum temperature for the column or cause breakdown of your sample. Where a sample contains components of a wide range of volatilities a single temperature (isothermal) may not be suitable for obtaining satisfactory separations of all the components. In such cases you have peaks which elute too rapidly and overlap each other, and ones which may take too long. In these cases it is usual to use a **temperature programme** which allows the chromatogram to start at a lower temperature, run through a period of controlled temperature rise (a ramp) and then remain at a chosen final temperature. This allows the initial part of the chromatogram to be stretched out and the final part to be compressed, which gives a more satisfactory overall trace.

The flow rate in GC is also easy to change, but not by a precisely known amount unless further time-consuming measurements are made. It has a comparatively small effect on the results. In contrast, temperature in liquid chromatography is less generally varied (it cannot be done as quickly as in GC, and it may have adverse effects on the flow in the column) but flow rate is easily varied and precisely controlled.

### 5.8.2 Composition of the mobile phase

In GC the identity of the carrier gas is not important, provided that it has suitable flow characteristics and is inert under the conditions used. In effect, this means that nitrogen is most usually used for packed columns and helium for capillary columns.

In liquid chromatography, however, the interactions between the molecules in the sample and those of the mobile phase are important and changes in mobile phase composition can dramatically change the separations that you get. Fortunately, it is easy to change the mobile-phase composition either by putting a new mixture in the reservoir (for a simple pump and isocratic system) or by changing the settings on a proportioning pump (isocratic and gradient). In either case, you must allow time for the column to equilibrate with the new mobile phase. Quite often you can do this by looking for a steady baseline. Most liquid chromatography is reverse phase (see Section 5.2.2) and so changes in mobile-phase composition should follow the rule:

- A less polar mobile phase gives quicker elution.
- A more polar mobile phase gives slower elution.

For normal-phase chromatography the rule is the opposite of this.

There will be many biological samples in which there is a variety of polarities and it may be impossible (or impracticable) to find a solvent mixture which will produce satisfactory separations for all the components, or it may just take too long. You can overcome this by using a **mobile-phase gradient** in which a starting composition is defined, followed by a programmed change to a final composition. For reverse-phase systems the gradient will be from more polar to less polar and the opposite for normal phase. As the chromatography proceeds, the more slowly moving components become subjected to the more strongly eluting mobile phase, so the whole chromatogram is compressed as compared with the isocratic method.

Compare this with the use of a temperature gradient in gas chromatography.

### 5.8.3 Changing the column

If none of the above changes produce the desired separation then a different column must be tried. For a skilled user this need not take too long but it is a more involved operation than any of those mentioned above. The system must be shut down (which involves cooling down for GC), joints disassembled, reassembled and checked for leaks. In GC the flow rate must be remeasured and the oven reheated.

### 5.8.4 Optimization

Unfortunately you cannot find the optimum performance for each of these variables one by one and then put them all together expecting to get the overall optimum. For example, the flow rate which is optimum for one mobile phase composition may not be optimum for another. However, when many similar samples are run, it is possible to home in on optimum conditions but it is time consuming. This may be offset by getting useful separations in the process even if they are not optimal. In the end, you may settle for the best that it is practicable rather than what is best by some absolute criterion.

There are now computerized methods which can search efficiently and automatically for optimum conditions. One such method is known as 'simplex optimization'.

## 5.9 Developments

One of the drawbacks of chromatography is that you need standards in order to identify components in your sample and, even then, you cannot be absolutely certain of the identification. In research laboratories you may well have samples for which you have no standards, especially if they contain novel substances. In contrast, spectroscopic techniques can give a lot of information about the structure and possible identity of unknown substances, often to such an extent that complete identification is possible. The two techniques are now frequently linked in order to combine their strengths. Spectroscopic methods that can scan a range of wavelengths very rapidly and are sensitive enough to respond to very small amounts of material are used as detectors. They give a spectrum of each component as it emerges from the column.

The first of such linkings was GC to mass spectroscopy (GC–MS) whereby the relative molecular mass of each component could be obtained along with structural information. More recently the very sensitive and rapid method of Fourier transform infrared (FTIR; see Section 3.3.5) has been used (to give GC–FTIR) so that an IR spectrum of each component is available which indicates what sorts of molecular structure are present in the component. The diode array UV detector coupled to HPLC has already been mentioned in this chapter (Section 5.5.3) and is now very widely used. In a similar manner HPLC and FTIR have been linked to give HPLC–FTIR and, although technically very difficult, HPLC can now be linked to MS (HPLC–MS). Such combinations are very powerful analytical systems and are providing analysts with more information about their sample at a faster rate than ever before.

These new combinations of techniques are sometimes referred to as 'hyphenated techniques'. You can see why from the initials.

## Further reading

Braun, R.D. (1987) *Introduction to Instrumental Analysis*, McGraw-Hill International Edition, ISBN 0 07 100147 6.
Chapters 24–26.

Harris, D.C. (1987) *Quantitative Chemical Analysis*, 2nd edn, W.H. Freeman, ISBN 0 716 71817 0.
Chapter 24 gives an overview of chromatography with some details of the instrumentation.

Skoog, D.A. and West, D.M. (1980) *Principles of Instrumental Analysis*, 2nd edn, Holt & Saunders International Edition, ISBN 4 8337 0003 4.
Chapters 24–26, however, surprisingly little on HPLC.

The ACOL 'teach yourself' style books include the following titles.

Lindsay, S. (1992) *High Performance Liquid Chromatography*, 2nd edn, Wiley, ISBN 0 471 93180 2.

Sewell, P.A. (1987) *Chromatographic Separations*, Wiley, ISBN 0 471 91371 5.
Emphasis on the theory, with some applications to GC and HPLC.

Willett, J. (1987) *Gas Chromatography*, Wiley, ISBN 0 471 91331 6.

For applications to food analysis, with some treatment of the theory, see:

Dickes, G.J. (1976) *Gas Chromatography in Food Analysis*, Butterworths, ISBN 0 50 033976 6.
Macrae, R. (1982) *HPLC in Food Analysis*, Academic Press, ISBN 0 50 022176 5.

# Electrophoresis

# 6

You may be wondering what electrophoresis means, and why it was not included in Chapter 4 on electrochemical techniques. The first part of the name obviously refers to an electrical effect, while the second comes from the Greek meaning movement. Thus we have movement brought about by electricity, but not the sort of movement that occurs in an electric motor; electrophoresis is the movement of electrically charged particles in some sort of suspending medium. What use can we make of electrophoresis? Like chromatography and centrifugation, it is a separation technique. Samples that are mixtures can be separated into their individual components. These, in turn, can be characterized and quantified, and even isolated and identified. Much work on protein structures has made use of electrophoresis and it is an important step in the process now known as DNA fingerprinting.

After reading through this chapter you should:

- know the principle of electrophoresis;
- be aware of factors affecting the separations;
- be acquainted with the more widely used methods;
- recognize the types of pattern produced and their interpretation.

## 6.1 General features of electrophoresis systems

### 6.1.1 The mechanism of electrophoresis

From the earliest times of experiments with electricity it has been recognized that a positive charge repels another positive charge but attracts a negative one. Similarly, a negative repels a negative but attracts a positive. It can be summed up as 'like charges repel, unlike charges attract'. The charges that we are concerned with are caused by an imbalance in the numbers of protons and electrons in molecular structures and on electrodes. The attractions and repulsions occur at a molecular level just as they do at a macro level. So, when charged molecules are free to move, as they are in a solution, or in a porous gel, applying a voltage by

We often call charged molecules 'ions' but only if they are relatively small. Large molecules that have many charges (both positive and negative, as you find with proteins and nucleic acids) are not usually referred to in this way, although we may well refer to their 'ionic charges'.

means of a pair of electrodes produces movement of the charged molecules.

The usefulness of electrophoresis lies in the finding that different molecules move at different rates, so a mixture will become separated into its various components as the movement proceeds. Thus we have a separation technique which produces results reminiscent of chromatography. However, do not confuse the two techniques; there are significant differences between them. Electrophoresis requires charged molecules in a conducting medium (usually a buffer in solution permeating some supporting porous material) and an applied voltage which creates an electric field. The buffer solution is not a mobile phase, although the ions in it move, as do the components of the sample. Thus positive ions (cations) and molecules move towards the cathode, while negative ions (anions) and molecules move towards the anode.

In molecules containing both positive and negative groups it is the **net** charge which determines the direction of movement. The numbers of charged groups depends on the pH of the buffer and, for each molecule, there will be a pH at which there is zero net charge. This is called the **isoelectric point, pI**. A molecule at its isoelectric point will not move in an electrophoresis system. At pH values less than its pI a molecule will have a net positive charge and hence move towards the cathode, while at pH values greater than its pI it will move towards the anode. Notice that the same molecule can be made to move in either direction depending on the pH.

### 6.1.2 Factors affecting the rate of movement

Now we must consider what factors affect the **rate** of movement of the charged particles. Only if the rates are different for different molecules can we obtain separations. The factors are fairly obvious if we recognize that particles of any size will be subject to frictional forces as they move through the buffer solution. Thus we can say that, other factors being equal:

- **smaller molecules move faster than larger ones;**
- **more highly charged molecules move faster than less charged ones.**

We can combine these two factors and say that the rate of movement is determined by the **charge to size ratio**. We have to be careful about what we mean by the size of molecules since in the presence of water the effective size may be enlarged by a strongly bound layer of water molecules. Very often size is taken to mean relative molecular mass and this is a reasonable approximation **provided that the molecules all have a similar type of structure**. However, even if they have similar relative masses, globular proteins may move at a different rate from fibrous ones and linear (chain-like) DNA molecules may move differently from circular (closed loop) ones. Thus we can also say that:

- the **shapes of molecules** will affect their rate of movement;
- the greater the **voltage** the faster the movement.

We have noted that the ions of the buffer will also be moving, positive in one direction, negative in the other, so they will be assisting the movement of molecules moving in the same direction and hindering those moving in the opposite direction. If the ions of the buffer are of similar size to each other these effects will largely cancel out but there is often some net effect, known as electroendosmotic flow, which can interfere with the separation process.

### 6.1.3 Electrophoresis cells

You may find it useful to refer to Figures 6.1 and 6.2 to see how the following general features are incorporated into particular systems.

The current that flows in electrophoresis produces heat because of the electrical resistance of the buffer. If the system gets too hot then parts may dry out or become distorted, and the movement of the charged molecules may be affected. For high-voltage systems and even for some normal-voltage systems, a cooling plate may be incorporated.

There is bound to be some electrolysis in an electrophoresis cell but it takes place at the electrodes. The system is designed so that the components of the sample do not reach the electrodes. However, electrolysis products from the buffer ions may affect the sample molecules in some way, the most likely being a change of pH near the electrodes. So the system is also designed to have a large volume of buffer surrounding the electrodes, and some sort of barrier to diffusion of these products to reduce the chance that they will reach the zone of movement of the sample components.

In early electrophoresis systems, in which only the buffer supported the sample, poor separations arose from convection and diffusion. The movement of the sample is more controlled if there is some sort of porous solid acting as a supporting medium. The various types have given rise to a variety of styles of electrophoresis system.

### 6.1.4 Types of supporting medium

*Paper (largely cellulose)*

Thick filter paper soaked in the buffer can act as a support medium but it still allows noticeable diffusion, so components appear as fuzzy spots and resolution is poor compared with other materials.

*Cellulose acetate*

This is a modified form of paper in which the hydroxyl groups of the cellulose have been acetylated. The resulting material is less hydrophilic than paper and so dries out rather easily and needs more care in handling. Its advantage is that it has a much more uniform pore structure than paper (the fibrous nature of the original cellulose is lost) so it allows much less diffusion. Consequently the results are clearer and better resolved.

*Gels*

An early material which showed the technique was feasible was **starch** but it has now been almost entirely replaced by others. **Agarose**, derived from seaweed extracts, is widely used in nucleic acid work. **Polyacrylamide** is a synthetic material having very suitable properties for protein analysis. Both agarose and polyacrylamide give extremely fine separations with the minimum of diffusion of the components. The pore size within these gels depends on the percentage of solid material in the gel and can be varied to suit different ranges of molecular sizes. It is even possible to produce gradients of density within the gel to extend the ranges of molecular sizes that can be analysed.

## 6.2 Typical electrophoresis apparatus and procedures

An electrophoresis system is housed in a container called a tank. Within the tank compartments are arranged so that the electro-

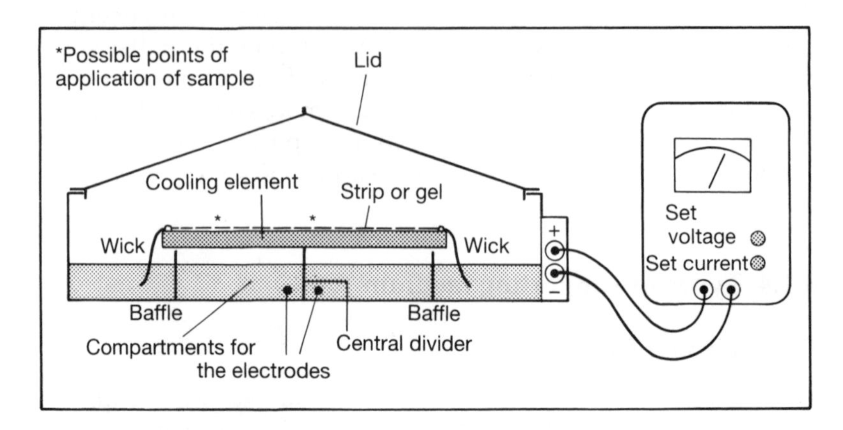


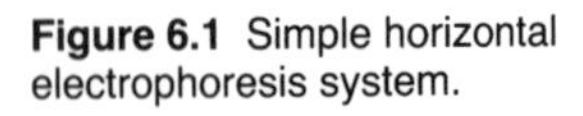
**Figure 6.1** Simple horizontal electrophoresis system.

phoresis occurs in either a vertical or a horizontal direction. In horizontal electrophoresis you can put the sample at any point along the support strip and movement of components may be in either direction (see Figure 6.1). However, you have to put the sample at the top of a vertical system and you must use a pH which ensures that all the components of the sample travel downwards (see Figure 6.2).

### 6.2.1 Horizontal systems

There are many variations in design: some tanks are larger (up to 30 cm × 30 cm), some smaller (mini-systems, perhaps 10 cm × 8 cm), but all follow a basic scheme which contains the following.

- **Central divider**: the two halves of the system each contain buffer and they should be at the same level to prevent syphoning from one to the other.
- **Compartments for the electrodes** which are shielded by baffles to retain electrolysis products but maintain electrical contact.
- **Electrodes** are usually of platinum wire.
- **Wicks** are soaked in buffer to produce good contact between the buffer and the **strip or gel** where the electrophoresis takes place, which is also soaked in buffer. In some cases the gel may even be submerged, e.g. DNA electrophoresis in 'submarine' agarose gels.
- **Point of application** of the sample may be near one end (if movement is all in the same direction) or in the middle (if movement in both directions is anticipated).
- **Lid** is peaked to allow condensation (from heating) to run to the sides and not drip onto the strip; also there will be some sort of switch allowing the voltage to be applied only when the lid is properly in place.
- **Cooling element**: this is included in high-voltage systems but may be omitted from normal-voltage systems.
- **Power supply** is highly stabilized, producing up to 300 V (normal range) or up to 1000 V (high-voltage systems). You can choose between constant voltage (preferred when there is direct contact between buffer and support medium), constant current (preferred when the contact is indirect or the gel is discontinuous; see below) or, in some cases, constant power (as in isoelectric focusing; see Section 6.3).

*Typical procedure*

1. **Sample treatment.** Try to arrange that the concentration of your sample is appropriate (often about 5 mg $cm^{-3}$ by the time all the reagents have been added); add buffer; add a marker,

i.e. a dye such as bromophenol blue which runs faster than any component in your sample, so that you can see the front of the movement; add sucrose or glycerol, if necessary, to make your sample dense for gel electrophoresis.

If you are unsure whether the concentration is adequate for showing the components you wish to see, you can put two or three loadings of each sample. Major components will show most clearly in the lightest loading, while trace components will be more visible in a heavier loading.

Then, once the tank is properly set up:

2. **Prepare the strip or gel.** If you are using cellulose acetate or paper, cut it to size; mark the proposed point of application with a pencil; soak the strip in the buffer, taking care that it is uniformly wetted, and blot it to avoid drops of buffer on the surface spoiling the movement of the sample.

   Gels have to be prepared beforehand and you need to have a recipe for this (see Box 6.1). While still liquid, they are poured into a tray and a plastic strip in the shape of a coarse comb is inserted to form wells for the samples. Once a gel has set, you carefully remove the comb and slide the gel out of the tray onto the platform in the tank. You can get ready-made gels commercially. They come in a plastic pouch, saturated with buffer and cast on a stiff plastic support which you can cut to size, if necessary.

   A period of, say, 5 min blank running will help to establish uniform conditions on the strip.
3. **Apply sample(s).** For strips or paper, put a spot or thin line of standards and samples on the same strip. Do this before the strip dries out (there are various methods which are best learnt from someone with experience). Position the strip and ensure it is in contact with the buffer (using the wicks, if appropriate).

   For gels, put a standard or a sample in each well.
4. **Make electrical connections.** DO NOT switch on until you have done all of the following: closed the lid; made sure the connections to the tank are the correct polarity; set both the voltage and current controls to their minimum positions and selected either constant voltage or constant current.

   Now switch on and increase the appropriate control until you have the required voltage or current.
5. **Run the sample(s).** The marker dye shows the furthest extent of the electrophoresis. When it has travelled a suitable distance, switch off and unplug at the mains.
6. **Staining and destaining.** You will need to have details of the solutions and times required for your particular method but, in general, you will place the strip(s) or the gels in a stain, ensuring complete wetting and probably allowing up to an hour for full staining. You then rinse the strips or gels to

Some of the materials and reagents used for preparing gels, particularly acrylamide, and for the staining, particularly ethidium bromide, and destaining are hazardous to health. You must know the procedures for handling them safely.

remove the bulk of the adhering stain and place them in a destaining solution to remove the stain from the background. Several changes of destaining solution over a period of a few hours up to a day may be required.

7. **Recording the results.** Cellulose acetate and paper can be dried but the colours may fade. Gels need special drying to avoid cracking and are not easy to store. You can, of course, photograph the patterns or draw a diagram of them. An alternative is to use a densitometer or an image analysis system to produce a graph of the colour intensity along the strip. DNA gels are usually stained with a fluorescent substance (ethidium bromide) and you need a special ultraviolet (UV) illuminator to see the bands. A typical set of results from horizontal electrophoresis is shown in Plate 7.

**Box 6.1 Procedures for preparing gels**

This is to give you an idea of what is done and you should not try it as though it were the full procedure. Some of the reagents are hazardous and you must use appropriate precautions.

**Agarose**

You need to decide what volume of gel solution you will use and then calculate the weight of agarose to give the final concentration you want. These concentrations are usually in the range 0.8–2%. For example, if you were preparing 100 $cm^3$ of a 1% gel you would need 1 g of agarose. The solvent is a buffer whose concentration and pH are dictated by the needs of the sample; 90 mmol $dm^{-3}$ 'Trisma', 90 mmol $dm^{-3}$ sodium borate with 2.5 mmol $dm^{-3}$ EDTA is typical.

The agarose is added to the buffer and the mixture is stirred and warmed (in a microwave if you wish!) until a clear solution is obtained. You then let it cool until it is 'hand-hot', pour it into a tray of a suitable size for your apparatus and allow the gel to solidify. It can then be transferred to the electrophoresis tank.

**Polyacrylamide**

Solutions needed:

- 30% acrylamide, the monomer, with a small amount of methylenebis-acrylamide for cross-linking;
- buffer, usually 'TRIS' pH 8.8;
- 10% ammonium persulphate, the polymerization initiator;
- 'TEMED', an amine to control the polymerization;
- (10% SDS if required).

Various volumes of these are mixed to give the required final percentage of acrylamide in the gel, such as 4% for stacking and 7.5%, 10% or 12% for the analytical gel. Polymerization is usually complete in about 1 h.

## 6.2.2 Vertical systems

The components of the system serve the same functions as in the horizontal system, but may look rather different. Vertical systems are used only for gels. It is important that the upper buffer

**Figure 6.2** A typical vertical electrophoresis system. (Courtesy Camlab, UK.)

compartment is well sealed against leaks. An earlier style had glass tubes containing the gel and, since the results showed up as thin discs of colour in the transparent rod of gel, the arrangement was often referred to as **disc gel electrophoresis**. It has now largely been replaced by **slab gel electrophoresis** although the slab is usually only about 1 mm thick and is supported between two glass plates. After staining, the results show as thin bands of colour.

The gel is formed in a holder which may be a pair of glass plates (10 cm × 10 cm or similar dimensions) separated by a spacer (1 mm thick), as shown in Figure 6.2. Frequently, a discontinuous gel system is used in which an extra layer of low-density gel is put on top of the analytical gel. This allows the samples to concentrate at the top of the analytical gel, resulting in a narrower band of material at the start and better resolution of the components. In either case, a serrated plastic strip (often called a comb) is placed at the top while the gel is forming. When the comb is removed, approximately 12 wells are left in the gel for receiving the samples.

Some of the procedures here are very similar to those for preparing gels for horizontal electrophoresis.

The samples have to be more dense than the electrophoresis buffer so that they stay in the wells once they have been added. This can be done by adding glycerol or sucrose to the sample. Neither substance consists of charged molecules so they do not interfere with the electrophoresis. As in horizontal electrophoresis, you may well add a coloured marker to run in front of any of the components so that you can see how far the electrophoresis has progressed.

The system is run with the gel supported by the glass plates. Some tanks are designed to run two gels at the same time so that a greater number of samples can be run at once. At the end of the run the gel has to be extracted and transferred to a staining solution. This is a delicate operation and you need to see it done and practise it to avoid breaking the gel. A sequence of staining and destaining is required as in horizontal electrophoresis.

*Typical procedure*

1. **Sample treatment.** Adjust the concentration of your sample so that its final value will be appropriate (such as $1\,mg\,cm^{-3}$ for proteins, or $0.25\,mg\,cm^{-3}$ for nucleic acids); add the buffer, glycerol or sucrose, the marker substance (e.g. bromophenol blue) and reducing agent such as dithiothreitol if it is important to avoid oxidation of the proteins.
   Then, once the tank is properly set up:
2. **Apply the samples.** Pipette suitable volumes of samples and standards into the wells. A plunger-operated pipette with a very fine tip is best for this. You may not be sure of the best volume to use for each sample, but you get over this by doing each sample in triplicate using, say, 5, 10 and $20\,mm^3$ ($\mu l$) volumes so that at least one of these is likely to give clearly defined bands that are not overloaded.
3. **Set the current.** Set up the power supply in the same way as for horizontal electrophoresis (making sure that the polarity of the electrodes will cause movement downwards). If you are using a discontinuous gel, set the current to, say, 20 mA per gel until the blue marker line reaches the junction of the gels. Then set the current to, say, 40 mA per gel until the marker dye is near the bottom of the gel. The actual currents used will depend on the dimensions of your gel.
4. **Stain and destain.** Switch off the power supply and disconnect it from the tank. Transfer the gel to the stain (watch a demonstration first!). Destain appropriately. A frequently used stain for proteins is Coomassie blue, which requires about 30–60 min followed by several soakings in dilute ethanoic (acetic) acid for the destaining over a period of hours. The result should be fine blue bands on a colourless background. Other stains are being

introduced, for instance a silver stain, which need more elaborate processing but can detect proteins at lower levels.

5. **Record the results.** Sketch, photograph, or use a densitometer or image analysis system to record your results. If you want to keep the actual gel, you will need to use a drier or a drying frame to prevent cracking and curling.

## 6.3 Variations and developments in electrophoresis

### 6.3.1 SDS-PAGE; the use of sodium dodecylsulphate in the polyacrylamide gel electrophoresis of proteins

Remember that denaturing means that the overall shape of the molecule and the particular twists and turns of the chain are disrupted but the chain is not broken (hydrolysed). The structure of SDS is $C_{12}H_{25}OSO_3^-$ $Na^+$. Its hydrocarbon portion associates with the hydrophobic parts of the protein structure, while its ionic group remains in contact with the surrounding water.

When the detergent sodium dodecylsulphate (SDS, sometimes called sodium laurylsulphate) is added to a sample, proteins become denatured and saturated with the detergent, which results in a fixed ratio of SDS to protein.

The ionic charges of the protein itself are effectively swamped by the negative charge on the detergent molecules, so the rate of movement during electrophoresis becomes dependent only on the size of the protein. The larger molecules move more slowly, the smaller ones faster. At the end of the electrophoresis we find that the distance moved by each protein is inversely related to the logarithm of its relative molecular mass ($M_R$). Thus we can use proteins of known $M_R$ to construct a calibration graph, and the $M_R$s of other proteins can be determined, as shown in Figure 6.3.

In all cases, you can make a quick visual estimation of how many sample components there are and roughly what their $M_R$s are, but for more precise work you should prepare a calibration graph. It is important to realize that you can only determine the $M_R$ of a sample protein, not its identity. Proteins of very different constitution may have similar $M_R$ values and hence travel almost the same distance during the electrophoresis.

If there is a very large range of $M_R$ within a sample, you can use gels with a gradient of density which allows the resolution of a wider range of $M_R$ within the same gel.

### 6.3.2 Isoelectric focusing

This is a another variation of the simple electrophoresis technique which makes use of a gradient. It may be carried out in a column or in a gel and uses special support material which is a mixture of ampholines. Ampholines have a variety of ionizing groups which, when subjected to an electric field, establish a pH gradient in the system. Then, wherever a sample is placed in the gradient, its components will move towards the position where the pH equals their isoelectric point, and there they stop. Even if a component is spread out at the start it will focus to a sharp band. The size of the protein is not important in this technique. Proteins have a variety of isoelectric points and so can be separated on this basis (see

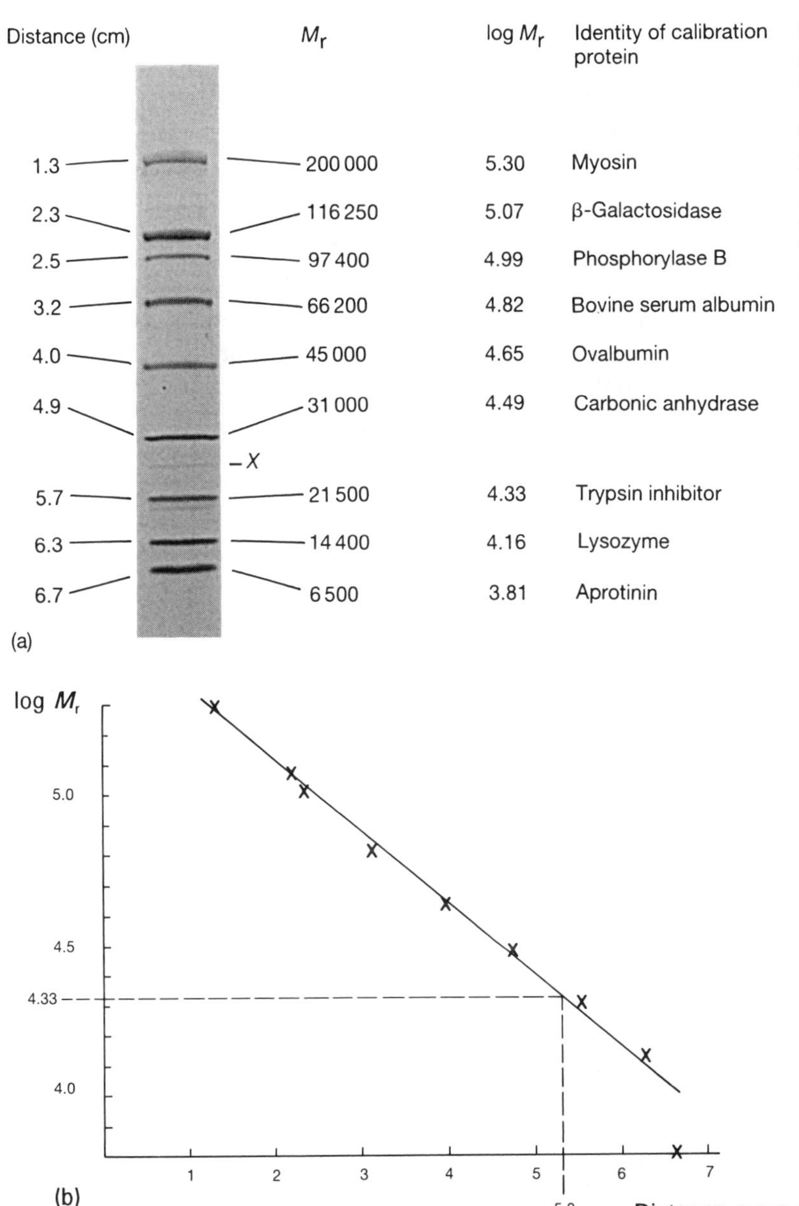


**Figure 6.3** Finding the relative molecular mass of a protein using SDS-PAGE. Proteins of known $M_R$ (a) are used to construct a calibration graph (b) from which the $M_R$s of other proteins can be determined. The minor band, *X*, in the standards may be a breakdown product of one of the larger proteins. You can see from the graph that log $M_R$ is 4.33, so the relative molecular mass of *X* is 24 500. (Courtesy of Bio-Rad Labs, UK.)

Plate 8). Nucleic acids do not, so the technique is not suitable for them.

### 6.3.3 pI determination

The pH gradient used in isoelectric focusing can also be used for the determination of the isoelectric point of a protein. If the electrophoresis is done at right-angles to the gradient with a line of sample along the gradient, the sample will move at different rates along the line and produce a curve. The point on the curve that has not moved at all (where the curve crosses the origin line) will be the pI.

### 6.3.4 Variable field electrophoresis

DNA presents a particular problem in electrophoresis because the molecules are long and thread-like. They have to be 'pulled' through the matrix of the gel rather like threading yarn through a sponge. Separation of very large sizes of DNA fragments, such as in the region of 1 million base pairs, is difficult but important these days because of work on genomes. Separations can be improved by reversing the field at programmed intervals (pulsed field), applying the field diagonally and alternately left and right to give a zig-zag progress or a combination of the two which produces a rotating field.

### 6.3.5 Immunoelectrophoresis

Instead of staining to detect protein components in gels you can use antibodies which react with the proteins. The antibody solution is applied after the electrophoresis and may be placed in wells or grooves cut into the gel near the line of the electrophoresis. It diffuses towards the proteins and gives visible lines of precipitation where it comes into contact with them. A distinctive pattern of arcs is produced showing the number and concentration of proteins. If a specific antibody is used a particular protein can be detected which might be an abnormal protein characteristic of a medical condition.

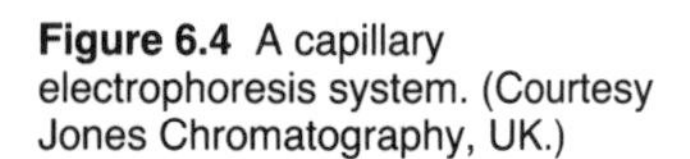

**Figure 6.4** A capillary electrophoresis system. (Courtesy Jones Chromatography, UK.)

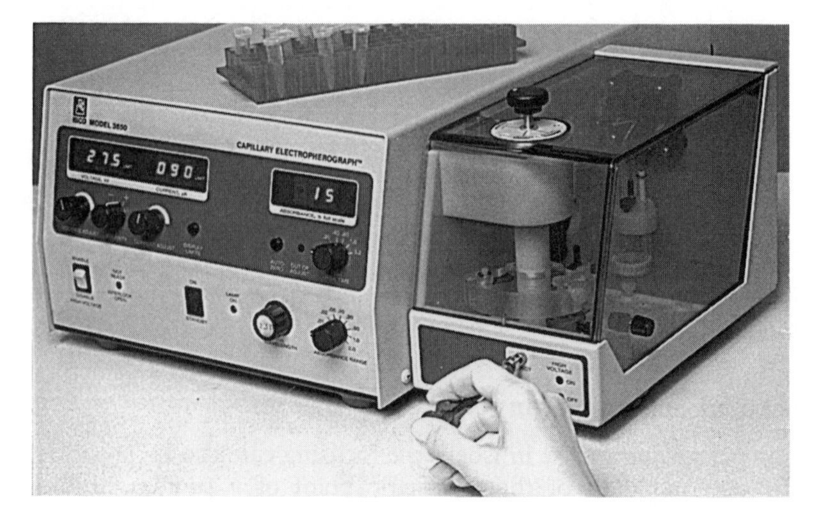

### 6.3.6 Blotting techniques

In nucleic acid analysis you often need to know more about the bands than a simple staining procedure will tell you. You will very probably want to know if a particular sequence of bases is present or if your fragments will bind with other fragments known as probes. Staining will prevent you from doing this. However, you cannot apply the probes directly onto the gel because they do not penetrate the gel structure (in contrast to the immunodiffusion mentioned in the previous technique). So the nucleic acids are removed from the gel by placing a membrane and layers of absorbent material on the gel. Thick paper tissue is very suitable. As the buffer is drawn out of the gel into the tissue it washes the nucleic acids onto the membrane and they can then be treated with probes to show whether there is binding or any other reaction.

The originator of the blotting technique was G. Southern, who transferred DNA fragments from agarose gels to nitrocellulose paper, and it became known as 'Southern blotting'. Since then variations of the technique have been devised both for DNA and for protein. They are called 'northern blotting' and 'western blotting' but not because of the names of their inventors!

### 6.3.7 Capillary electrophoresis

In this technique there is no support material. The sample travels inside a capillary tube and is subject to effects at the inside surface of the tube similar to the endosmotic flow mentioned above (Section 6.1.2). The elution of the components from the end of the capillary is monitored in a similar manner to HPLC (Section 5.5). Samples can be extremely small and results can be obtained in minutes, compared with at least an hour in conventional electrophoresis. The disadvantage is the price, currently around £20 000, but as with so many instrumental techniques this is likely to decrease as the technique becomes more widespread, and it may even become as common as HPLC. A typical capillary electrophoresis instrument is shown in Figure 6.4.

## Further reading

The ACOL 'teach yourself' style books include the following title.

Melvin, M. (1987) *Electrophoresis*, Wiley, ISBN 0 471 91374 X.

A useful survey with some detail of theory is:

Simpson, C.F. and Whittaker, M. (eds) (1983) *Electrophoretic Techniques*, Academic Press, ISBN 0 50 041852 6.

# 7 Radiochemical techniques

Radiochemical techniques are invaluable in biological studies, since you can make measurements at very low levels and you can often avoid complex extraction procedures. Another particular advantage is that radioactive versions of biological molecules follow exactly the same reactions as the non-radioactive ones, so metabolic pathways can be determined precisely. This is not always the case with other methods of investigating them. Furthermore, because the radiation affects photographic film you can literally produce pictures of the distribution of radioactivity in tissues and even whole organisms, so you can show the location of particular molecules in cells and organs.

Today, however, we are much more conscious of the hazards of the overuse of radioactive materials and other techniques such as fluorescence microscopy, immunological methods and stable isotope techniques have been developed which provide similar information at a similar level of sensitivity. Nevertheless, radiochemical methods will continue to provide a useful and efficient tool in the analysis of biological materials.

After working through this chapter you should:

- know the characteristics of the common subatomic particles;
- know the characteristics of the common radioactive emissions;
- recognize how half-life and energy of emission affect the use of radioisotopes;
- be able to choose appropriate methods of detection;
- be able to perform calculations using count rate, efficiency, activity and specific activity.

## 7.1 The origin and nature of radioactivity

Why are some substances radioactive and others not? And what is radioactivity anyway? The answer to the first question is found in the nuclei of atoms. When you are dealing with the chemistry

**Table 7.1** Characteristics of subatomic particles

| *Name* | *Symbol* | *Approximate relative mass*[a] | *Relative charge*[a] | *Comments* |
|---|---|---|---|---|
| Proton | p | 1 | +1 | Found in the nucleus |
| Neutron | n | 1 | 0 | Found in the nucleus |
| Electron | $e^-$ | 0.00054 | −1 | $\beta^-$-radiation[b] |
| Positron | $e^+$ | 0.00054 | +1 | $\beta^+$-radiation[b] |
| Alpha-particle | $\alpha$ | 4 | +2 | $\alpha$-radiation[b] |

[a] Relative to the proton.
[b] When emitted from the nucleus.

and biochemistry of atoms and molecules you are studying the behaviour of their outermost electrons but now we are transferring our attention to the heart of the atom. There are only certain combinations of protons and neutrons that give stable nuclei. Those that are unstable exist for a while, sometimes for a very long while, but eventually undergo a transformation and emit some form of radiation, or radioactive emission, that can be detected at a distance from its source.

Nuclear transformations are often called 'disintegrations' but this gives a false impression of the process. The transformations that we are considering leave the nucleus largely intact.

There are three sorts of radioactive emission, which have been labelled alpha ($\alpha$), beta ($\beta$) and gamma ($\gamma$) just because they were discovered in that order. Each of them causes some ionization of the material through which it is passing, so they are often collectively known as **ionizing radiation**. Of these, only $\gamma$-emission is electromagnetic radiation; the other two are emissions of subatomic particles. Table 7.1 lists the nature and characteristics of the common subatomic particles that you need to know at this stage.

### 7.1.1 Isotopes

All the atoms in an element are chemically identical but they do not all have exactly the same atomic structure. The number of protons and electrons is always the same but the number of neutrons in the nucleus may vary. These different forms of the atom are known as **isotopes** and they are distinguished from each other by their mass. You should note that the term isotope does not imply radioactivity; some isotopes are radioactive and some are stable. Those that are unstable may have too few neutrons, too many neutrons or too large a nucleus.

We refer to isotopes in terms of the element name and the mass number. Two radioactive isotopes widely used in biological experiments are carbon-14 and phosphorus-32, while carbon-13 is a stable isotope also used in biological work. The only exception to the rule about naming is that the isotopes of hydrogen have

**Table 7.2** Radioisotopes commonly used in biological work

| *Name* | *Symbol* | *Type of emission* | *Half-life* |
|---|---|---|---|
| Tritium | $^{3}H$ | Soft $\beta^-$ | 12.26 years |
| Carbon-14 | $^{14}C$ | Soft $\beta^-$ | 5730 years |
| Sodium-22 | $^{22}Na$ | Mainly $\beta^+$ | 2.6 years |
| Sodium-24 | $^{24}Na$ | Hard $\beta^-$ | 15 h |
| Phosphorus-32 | $^{32}P$ | Hard $\beta^-$ | 14.3 days |
| Sulphur-35 | $^{35}S$ | Soft $\beta^-$ | 87 days |
| Chlorine-36 | $^{36}Cl$ | Hard $\beta^-$ | 300 000 years |
| Potassium-42 | $^{42}K$ | Hard $\beta^-$ | 12.4 h |
| Calcium-45 | $^{45}Ca$ | Soft $\beta^-$ | 165 days |
| Iron-55 | $^{55}Fe$ | $\gamma$ | 2.7 years |
| Iron-59 | $^{59}Fe$ | Soft $\beta^-$ | 45 days |
| Iodine-125 | $^{125}I$ | $\gamma$ | 60 days |
| Iodine-131 | $^{131}I$ | Hard $\beta^-$ | 8 days |

The name isotope means 'same place' and was given because all the isotopes of an element must be in the same place in the periodic table. Another term, 'nuclide', seems to mean the same but strictly refers to any atom of specified nuclear composition without implying a series with the same atomic number.

special names: deuterium is 'hydrogen-2' (a stable isotope) and tritium is 'hydrogen-3' (a radioactive isotope).

We also refer to isotopes by chemical symbols with added superscripts (the mass number, i.e. the number of protons plus the number of neutrons) and subscripts (the atomic number, which is the number of protons). Thus, for the carbon and phosphorus isotopes mentioned we have:

$$^{13}_{6}C \quad ^{14}_{6}C \quad ^{32}_{15}P$$

but usually we omit the subscripts since we are more concerned with the identity of the element rather than its atomic number. Again, the isotopes of hydrogen are an exception and there are special symbols for them, D and T, although the symbols $^{2}_{1}H$ and $^{3}_{1}H$ are also used. Table 7.2 gives a list of some radioisotopes commonly used in biological work.

In older texts the super- and subscripts may appear to the right of the symbol.

### 7.1.2 The nature of radioactive emissions

The brief indications of damage to tissue are only the beginnings of an indication to health hazards. This is an extremely complex topic; it has to take into account the time and intensity of the exposure, the type of radiation and the sensitivity of the tissue to that radiation. It is probably best to regard all radiation as potentially hazardous and ensure that you work to the highest safety standards.

**Alpha emission** is mostly found with atoms whose nuclei are too large for stability, i.e. polonium-210 and beyond. It consists of $\alpha$-particles (see Table 7.1) which are, in effect, helium nuclei. They interact strongly with any material they pass through because, on a subatomic scale, they have a high mass (4) and charge (+2). They rapidly acquire two electrons and become helium atoms. They travel very short distances only but have very high energies and do a lot of damage to tissues. However, it is easy to shield against this type of emission since any thin material will absorb it (see Table 7.3).

**Beta emissicn** comes in two sorts. The first, beta-negative ($\beta^-$),

**Table 7.3** Characteristics of radioactive emissions

| *Type* | *Penetration (approx.) through* | | | | *Health hazard* | *Shielding materials* |
|---|---|---|---|---|---|---|
| | *Air* | *Water* | *Al* | *Pb* | | |
| Alpha | 1 cm | 1 mm | 0.1 mm | 0.05 mm | High on actual contact | Any thin material |
| Beta | | | | | | |
| Soft | 10 cm | 1 cm | 1 mm | 0.5 mm | Low to moderate | Lab. glassware, perspex |
| Hard | 50 cm | 5 cm | 1 cm | 0.5 cm | Moderate to high | Perspex, thick Al |
| Gamma | 25 m | 5 m | 20 cm | 10 cm | Low to moderate | Lead bricks, concrete blocks |

There are no electrons or positrons as such in the nucleus but they are produced and ejected during the nuclear transformation; a neutron may turn into a proton and an electron, or a proton may turn into a neutron and a positron. Other particles, nutrinos, are also emitted but are not detected by ordinary laboratory apparatus and cannot be used as analytical tools.

is high-energy electrons from nuclei with too many neutrons, such as carbon-14 and phosphorus-32. We distinguish between soft (less energetic, such as carbon-14) and hard (more energetic, such as phosphorus-32) if we need to indicate their penetration characteristics, as we might in order to suggest which types of detection and shielding would be appropriate. The second, beta-positive ($\beta^+$), is the emission of positrons from nuclei with too few neutrons such as sodium-22. Positrons are the same as electrons in mass but have opposite charge. They are, thus, 'anti-electrons' and undergo annihilation when they interact with electrons, giving a characteristic gamma ($\gamma$) radiation.

Whether positive or negative, $\beta$-radiation interacts less strongly with matter than $\alpha$- because of its lower mass and single charge. The energy that it loses as it passes through tissue is thus distributed through a greater volume and is less likely to cause serious damage to cells.

**Gamma ($\gamma$)-radiation** is electromagnetic radiation, as we mentioned when discussing the electromagnetic spectrum in Section 3.2.1. It does not consist of particles that have mass or charge, although we sometimes refer to gamma-ray photons as 'particles'. It is more energetic than X-rays and more penetrating. It interacts even less with matter than $\alpha$- or $\beta$- and so is lowest on a rough scale of damage to tissue.

Energies of radioactive emissions are usually referred to using the scale of electron-volts (eV), i.e. the energy equivalent to that acquired by an electron when it has been accelerated by 1 V. One electron-volt is equivalent to $1.6 \times 10^{-19}$ J. A typical energy for an emission from carbon-14 is 0.1 MeV (100 000 eV) whereas the energy required to break a carbon–hydrogen bond is approximately 2 eV, so there is the potential to break 50 000 C—H bonds! Of course, this is not what happens; much of the energy gets converted into vibrations and electronic excitation without breaking bonds but, even so, one emission does break a relatively large number of bonds. However, for the normal scale of working in a laboratory, the amount of decomposition resulting from this is virtually undetectable chemically.

## 7.2 Units for measuring radioactivity

### 7.2.1 Count rate

When you use a Geiger counter or any other radioactivity detector you are counting the emissions from the radioactive source that have reached the detector. However, it is the rate of counting that is important, so the number of counts is divided by the duration of the measurement to give the count rate. Most often this will be in **counts per minute (cpm)** but other units are possible. Detectors may give the total count, a direct read-out of the rate or both. For simple measurements you may only need to compare count rates of various samples but you must make sure that they are measured under identical conditions in order to get the same efficiency of counting (see below). Also, you must count for a sufficient time to get a count rate that is statistically valid (see Box 7.1).

**Box 7.1 Statistics of counting**

The emission of radiation from radioactive atoms is a random process. We cannot predict exactly when a particular atom will undergo a nuclear transformation but we can say that, over a certain period of time, there will be a certain proportion of transformations. In practice this means that, for repeated measurements, we do not get exactly the same count each time. Sometimes the variation is surprisingly wide but it is purely due to the random nature of radioactivity. So we must apply statistical methods to our handling of data from radioactive experiments. These methods recognize that we can only obtain high precision if we use large numbers of measurements. Alternatively, we may try to ensure that the number of counts in each measurement is sufficiently large for the statistical precision to be compatible with the overall precision of the experiment.

Since we can only measure **whole** numbers of counts per minute and we cannot have negative numbers of counts the appropriate distribution for repeated measurements of a sample is the **Poisson distribution**. At high levels of counting this distribution is very similar to the normal distribution discussed in Box 1.3 but at low levels it differs significantly. However, at all levels the Poisson distribution has the characteristic feature that the standard deviation of a series of measurements is equal to the square root of the mean. We can use this to estimate the precision of a **single** measurement and we find that a higher count (regardless of the time taken) has a higher precision, as shown by a smaller relative standard deviation.

| Number of counts ($N$) | Standard deviation $s = \sqrt{N}$ | Relative std dev. $r = s/N$ | Percentage precision (%) |
|---|---|---|---|
| (values quoted to two significant figures) | | | |
| 100 | 10 | 0.1 | 10 |
| 1 000 | 32 | 0.032 | 3.2 |
| **10 000** | **100** | **0.01** | **1** |
| 100 000 | 320 | 0.0032 | 0.32 |

You can see that if you ensure that the total count in an experiment is at least 10 000 counts you will have a precision of counting similar to that sort of experimental precision (1% or better) you would expect to obtain in your manual operations during the experiment. For lower counts than this you must recognize the lower precision, however good your technique is.

### 7.2.2 Background

Even when your detector has no sample you will find that it is giving a certain count rate. This is the background radiation that reaches us from outer space (cosmic rays) and the natural levels of radioactivity in rocks containing isotopes such as potassium-40, uranium-238 as well as the gas radon-222. Thus, whenever you make measurements of radioactivity you take readings to establish the background count rate before you measure your samples. You must subtract the background from your sample count rate.

Of course, you try to ensure that your detector is shielded as much as is practicable from background radiation and hence gives as low a background count as possible.

Suppose you expose a plant to radioactive carbon dioxide. After some time you remove the plant from that atmosphere and combust some of the leaves. The carbon dioxide released by the combustion gives a count rate of 3650 cpm. A standard known to produce 1 mg of carbon dioxide gives a count rate of 2450 cpm. For each of these measurements the background was 50 cpm. You can then calculate that the carbon dioxide that had been taken up by the plant was:

$$\frac{3650 - 50}{2450 - 50} \times 1\,\text{mg} = 1.5\,\text{mg}$$

### 7.2.3 Activity

There are times when you have to know the actual rate of nuclear transformation (disintegration) in your sample. You may be comparing results from one experiment to another or publishing a report. If you have bought some radioactive material for an experiment you need to know the amount of radioactivity it contains. In a metabolism experiment there might be two metabolites and you wish to find out which is the more important metabolic pathway. The quantity you use for all these situations is the **activity**. It is defined as the **number of disintegrations per second (dps)**. The SI unit for this is the **becquerel (Bq)**, so an activity of 1 Bq means that there is one disintegration per second in the sample. This is a very low activity and most often you would be using sample with activities measured in kilobecquerels (kBq), while samples bought for elaborate investigations might have activities of megabecquerels (MBq).

An earlier unit of activity was the curie (Ci), which was $3.7 \times 10^{10}$ dps. This is a very large unit and so your results would usually be in mCi and $\mu$Ci. You may still come across these units is older books.
Henri Becquerel (1852–1908) was the discoverer of radioactivity. He found that an unexposed photographic plate had become 'fogged' by some uranium minerals left in the same drawer. The isolation and identification of the elements responsible for the effect was the painstaking work of the Curies (Marie, 1867–1934, and Pierre, whom she married in 1894). The story of these discoveries is well worth reading.

### 7.2.4 Specific activity

Not all the atoms in a sample are radioactive; sometimes there will be many and the sample will be highly radioactive, other times there will be few and the activity is low. It is, of course, quite possible to mix an active sample with an inactive sample of the same substance and, in effect, dilute the activity. The same total amount will be there but it will be in a greater amount of material. When we take into account both the activity and the mass of the material we refer to the **specific activity** and measure it in becquerels per milligram ($Bq\,mg^{-1}$) or kilobecquerels per millimole ($kBq\ mmol^{-1}$) or any similar units. Some types of analysis measure only a portion of the total activity and use the specific activity in a similar manner to using concentrations.

### 7.2.5 Efficiency of counting

You can detect radioactive transformations only if the emitted radiation enters the detector and causes a response. Radiation that has been emitted in a direction away from the detector cannot be reflected back into the detector in the way that light can and so much of the radiation may not be counted. Additionally, radiation travelling towards the detector may be absorbed by the materials it travels through, including the walls of the detector itself. In contrast, highly penetrating $\gamma$-radiation may pass through a detector without triggering a response and so, again, some of the activity is not detected. In other words, the efficiency of the detection is always less than 100% and sometimes very much less.

For experiments where it is important to know the activity of the sample you need to be able to calculate the efficiency. It may vary from one experiment to another, perhaps because of differences in the position of your sample, its thickness or the nature of other substances in the sample, although you would expect to take precautions to minimize these. In simple measuring systems you just find the count rate of a sample of known activity. Suppose you have a sample of sodium carbonate whose activity is 1 kBq and you find it gives 600 cpm (after subtraction of the background) in your detector. Then you can calculate that:

$$\text{since } 1\,\text{kBq} = 1000\,\text{dps, there are } 1000 \times 60 = 60\,000 \text{ disintegrations per minute (dpm).}$$

$$\text{The efficiency} = \frac{\text{count rate (cpm)}}{\text{disintegrations per minute (dpm)}} \times 100\% = \frac{600}{60\,000} \times 100\% = 1\%$$

Such a figure is typical of a Geiger counter. Other systems can give much higher efficiencies and some have other methods of determining efficiency, even automatic ones. (Table 7.4 includes some typical efficiency figures.) Once you know the efficiency of your system you can determine activities and specific activities of test samples. For example, a count rate of 480 cpm (after subtrac-

**Table 7.4** Summary of detection methods

| *Method* | *Applications* | *Typical efficiency* | *Comments* |
|---|---|---|---|
| Geiger counter | $\alpha$ | Variable | |
| | Soft $\beta$ | Less than 1% | Tritium not detected unless open window tube used |
| | Hard $\beta$ | 1–5% | |
| | $\gamma$ | Very low | |
| Solid scintillation | $\gamma$ | Up to 60% | |
| Liquid scintillation | Soft $\beta$ | Up to 50% for tritium<br>Up to 95% for carbon-14 | |

tion of the background) measured at 20% efficiency corresponds to an activity calculated thus.

First convert your count rate (cpm) to disintegrations per minute (dpm):

$$\text{count rate (cpm)} \times \frac{100\%}{\text{efficiency }\%} = 480 \times \frac{100\%}{20\%} = 2400\,\text{dpm}$$

This has to be converted to Bq (dps) by dividing it by 60:

$$\text{activity} = 2400/60 = 40\,\text{Bq}$$

If this were found for a plant sample which contained 2 mg of a pesticide, its specific activity would be:

$$40\,\text{Bq}/2\,\text{mg} = 20\,\text{Bq}\,\text{mg}^{-1}$$

From this you might calculate the proportion of the administered dose of the pesticide that had been taken up by the plant.

### 7.2.6 Half-lives and rates of decay

Most of the commonly used radioisotopes decay into stable isotopes. Thus, once a radioactive atom has undergone a nuclear transformation it is no longer radioactive. The activity of a sample depends on the number of radioactive atoms in the sample and so it decreases over a period of time. The rate of decrease is a characteristic of the isotope and is not affected by external conditions such as temperature and pressure. We most usually refer to this rate in terms of the **half-life**, which is defined as the time taken for the activity of a sample to decay to half the value at the start of the measurements. It does not matter what the activity of the sample is or when you started the measurements, it always takes the same time to fall to half. The decay thus follows the curve shown in Figure 7.1.

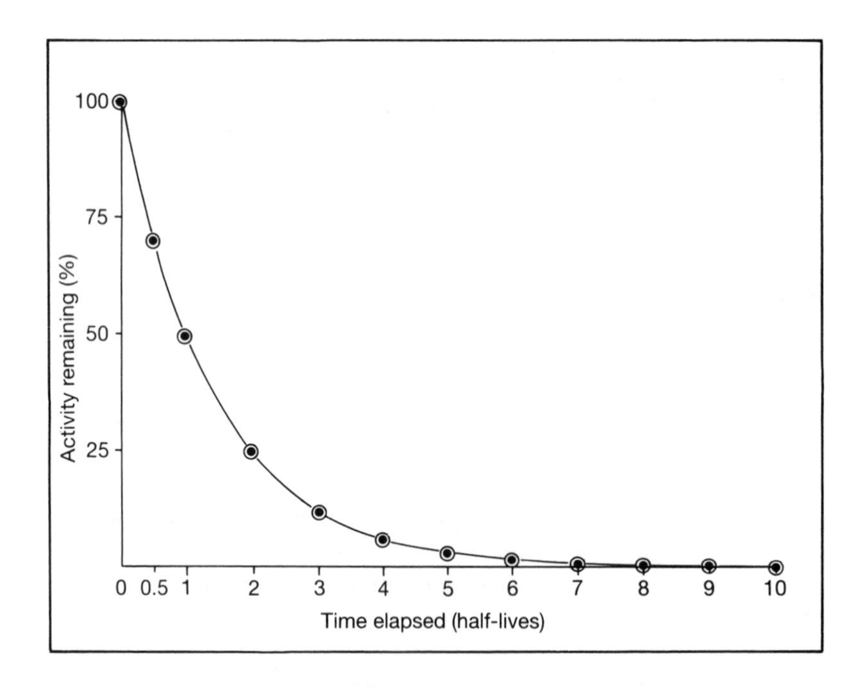


**Figure 7.1** Decay curve of a radioactive nuclide.

If you plot the logarithm of the per cent remaining activity on the *y*-axis you get a straight-line graph which can be used for the accurate determination of the half-life of an isotope.

Theoretically, then, the activity never quite falls to zero and the full life-time of an isotope is effectively infinite. In practice we need to know the progress of the decay over a finite time and the half-life is a convenient measure of this. Lengths of half-lives vary enormously from one isotope to another and you can see some examples in Table 7.2. For short-lived isotopes such as phosphorus-32 (14.3 days) and sodium-24 (15 h) the decay is significant from day to day and has to be taken into account in the measurements. For longer-lived isotopes such as sulphur-35 (87 days) and tritium (12.3 years) the decay will be significant only if the experimental measurements take place on a similar time scale. The decay of the long-lived isotopes, such as carbon-14 (5730 years) and chlorine-36 (300 000 years), is insignificant over the period of any ordinary experiment, although the technique of radiocarbon dating does detect the decay over periods of a few hundred to a few thousand years.

*Calculating present activities*

If we need to take the decay into account there is a simple formula for determining how much the activity has decayed over a period of time. We can use the symbols $A_t$ for the present activity, which we wish to calculate, and $A_0$ for the activity originally; $t$ for the time elapsed and $t_h$ for the half-life. Then:

$$A_t = A_0 \times (0.5)^{(t/t_h)}$$

Suppose you had started an experiment using 500 kBq of phosphorus-32 one week ago. You could have applied this as phosphate to the roots of a plant and you might now be measuring the proportion that had been taken up by the plant. You could measure the activity in the plant now and compare it with the total activity of the phosphorus-32 now. So, because of the half-life ($t_h$ = 14.3 days), the original activity ($A_0$ = 500 kBq) has decayed in a period of one week ($t$ = 7 days).

You would calculate, to three significant of figures, that:

$$t/t_h = 7/14.3 = 0.490$$

so $$(0.5)^{0.490} = 0.712$$

and $$A_t = 500\,\text{kBq} \times 0.712 = 356\,\text{kBq}$$

It is this last figure that is the present total activity.

*Half-lives and disposal methods*

The half-life also has a bearing on the methods of disposal of radioisotopes. Short half-life isotopes can be stored until the activity is sufficiently low for them to be regarded as low-level waste. For small-scale laboratory samples this is often taken to be about seven times the half-life since, by that time, the activity will be less than 1% of the original (you can see this from the graph in Figure 7.1). For long half-life isotopes this cannot be done since the activity does not diminish sufficiently over a period of a few years. These have become an environmental problem since burning or dumping at sea does not destroy the activity, it only disperses it, and indefinite storage underground may still give rise to leaching of the activity into watercourses. As yet, there is no general agreement as to the long-term safety of any method.

## 7.3 Methods of detecting radioactivity

We have described the nature of the main types of radioactive emission in Section 7.1 and have indicated that they have high energies since they cause many hundreds or even thousands of ionizations. The Geiger counter makes use of this ionization effect. Alternatively, the energy of the radiation may be used to produce flashes of light from scintillator molecules and this is used in scintillation counters.

### 7.3.1 Geiger counters

The basis of the best-known radiation detector is a tube, the Geiger tube, which is often a metal cylinder about 10 cm long and

Hans Geiger (1882–1945) was a German physicist who worked with Rutherford in Manchester during the early years of this century. They determined the nature of $\alpha$-particles and devised an $\alpha$-particle counter. In the 1920s Geiger worked with Muller at Keil and developed the present instrument from the original counter.

about 5 cm in diameter. One end of the tube has a thin material such as mica or a plastic film forming an end-window. The radiation can pass through this window without being strongly absorbed. In the centre of the tube is an anode maintained at about 400 V by a power supply. The cathode is incorporated into the wall of the tube. There is a low pressure of argon, with a small proportion of bromine or methane, in the tube. There are, however, many variations in the style of construction to give better performance for particular samples (solids, liquids, surfaces or others) and for particular types of radiation ($\alpha$, $\beta$ or $\gamma$). Figure 7.2 shows some of these.

When a radioactive emission enters the tube it produces electrons and positive ions from the argon. The electrons accelerate towards the anode and produce more ionizations in the process, resulting in an avalanche of electrons reaching the anode. This becomes the electrical pulse that is registered as one count. There is then a brief period in which the anode has to regain its voltage and is unresponsive (the dead-time). At very high count rates this dead-time becomes significant and you have to apply a correction formula to the observed count rate, but you are unlikely to encounter this in routine work.

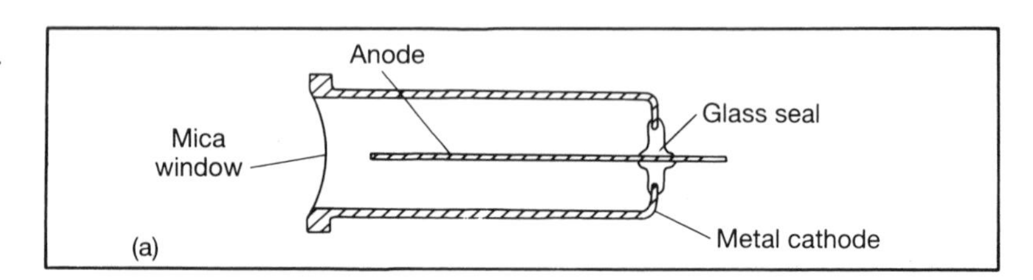


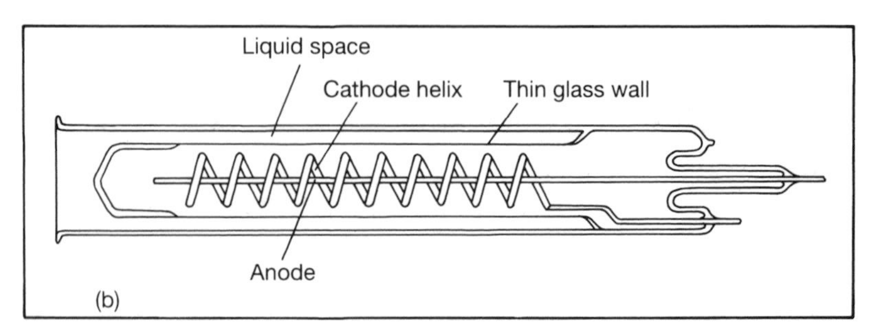


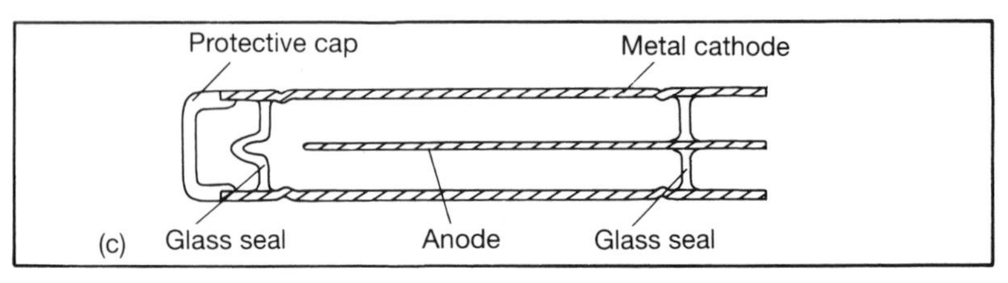


**Figure 7.2** Some typical Geiger tubes: (a) mica end-window alpha or beta tube; (b) Liquid counter beta tube; (c) metal-walled gamma tube.

A Geiger counter is rugged, simple to use, portable and cheap. You can easily do field-work with it and it is particularly suitable for monitoring surfaces for contamination. However, it is not an efficient detector in the sense disussed in Section 7.2. This does not make it inaccurate for quantitative work, just less able to detect low levels of radioactivity. Soft $\beta$-radiation, such as from carbon-14 and sulphur-35, is substantially absorbed by the end-window and so its efficiency of detection is very low indeed, perhaps 0.1%, depending on the set-up. In the case of tritium the $\beta$-radiation does not penetrate even the thinnest window and it is not detected. Some open-window Geiger tubes are used to overcome these low efficiencies but they require a continuous supply of gas and are not convenient for general-purpose work. Scanners for radioactive TLC plates use this type of Geiger tube. There is also a very low efficiency in the detection of $\gamma$-radiation because it is highly penetrating and passes through the gas in the tube, producing hardly any ionization. Finally, a Geiger tube in normal use does not distinguish between different radiations and only gives the total count of the radiation that it has detected.

### 7.3.2 Scintillation counters

Some substances detect radioactivity by producing flashes of light as they re-emit the energy acquired from the radioactive emission. These flashes are detected with high sensitivity by a photomultiplier tube.

Photomultipliers are also used as the light detector in sensitive spectrophotometers.

Such a system is very fast, which gives it an advantage over a Geiger tube in counting higher activities. Also, the intensity of the flash can be monitored which allows the energy of the radiation to be determined, so the system can discriminate between different radioisotopes. However, most scintillation systems are not portable and are not designed for monitoring surfaces. They are more likely to be highly automated and large numbers of samples can be measured (overnight, without supervision if necessary), leaving the operator to do less routine tasks. There are two main sorts of scintillation system, both of which are designed to overcome the limitations of the Geiger tube.

*Solid scintillation*

You need a dense material rather than a gas to detect $\gamma$-radiation. Sodium iodide doped with thallium iodide is generally used. It is often formed in the shape of a well so that the sample is surrounded as much as possible to give greater counting efficiency. Figure 7.3 shows one such arrangement but other designs achieve the same purpose.

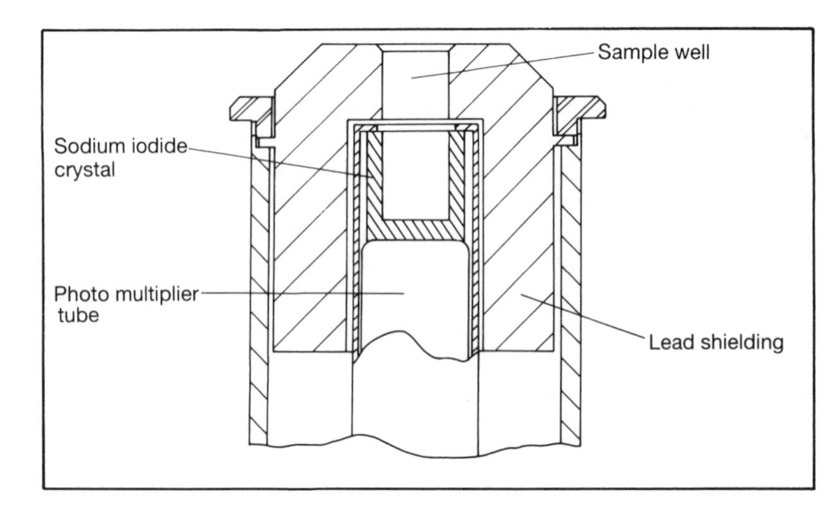


**Figure 7.3** Cross-section of a solid scintillation counter.

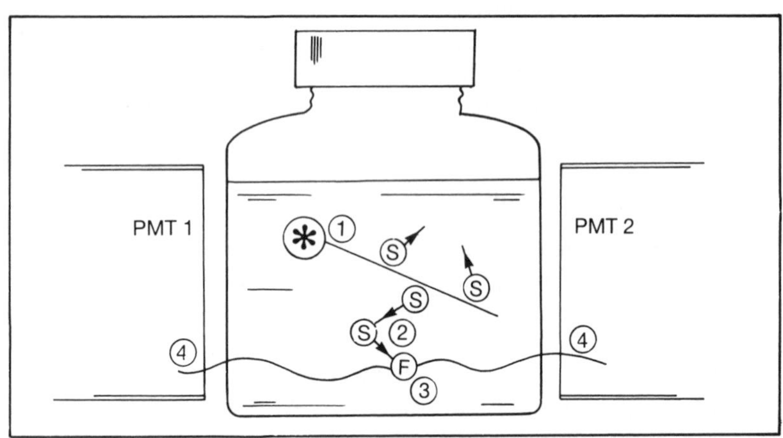


**Figure 7.4** The liquid scintillation process. (1) Radioactive emission energizes solvent molecules (S). (2) Energy passed via solvent to fluor (F) (scintillator). Chemical quenching can occur when species such as $O_2$ and $H^+$ are present. They do not pass on the energy efficiently. (3) Emission of a flash of light; wavelength approx. 400 nm. Colour quenching can occur when the colour of the sample is such that it absorbs this wavelength. (4) Simultaneous detection of the flash by both photomultiplier tubes, PMT 1 and PMT 2, triggers the counting system. Notes: Quenching reduces the efficiency of the system.
(S) The solvent is usually methylbenzene or similar compounds which are best for process 2.
(F) A widely used fluor is 'POP', which is diphenyloxazole,

*Liquid scintillation*

This system was designed particularly for soft $\beta$-emitters such as carbon-14, tritium and sulphur-35, and it is unique in having the sample mixed in with the scintillator material (see Box 7.2). The radiation does not therefore have to penetrate through any walls of containers, so there is no absorption hindering the detection. The system therefore offers the possibility of 100% efficiency but it is reduced by other factors.

Another term for scintillator is 'fluor'.

*Quenching*

Anything that interferes with the transfer of energy through the solvent to the scintillator molecules or with the transmission of the light to the photomultiplier tube is said to cause **quenching**. It makes the detector register a lower energy for the radiation and

**Box 7.2 Scintillator cocktails**

In liquid scintillation mixtures, the best solvents are methylbenzene (toluene), dimethylbenzene (xylene) and trimethylbenzene, for two reasons. First, the scintillator molecules, diphenyloxazole ('POP') and others which are included sometimes to improve the efficiency of the system, are soluble in such non-polar solvents rather than water. Second, they pass on the energy from one molecule to another very efficiently. However, the majority of biological samples are aqueous and you have to include a detergent in these samples to ensure that you have a single phase, otherwise counting efficiency is poor and unreproducible. The detergent does interfere with the scintillation process, reducing the efficiency (quenching). You do not usually expect to recover your sample from the scintillator mixture after the counting and the solvents are somewhat toxic, so the laboratory has to have proper disposal measures.

reduces the counting efficiency. There are two main causes of quenching. One, which prevents the efficient transfer of energy from one solvent molecule to another and finally to the scintillator molecules, is known as **chemical quenching** and can occur when substances such as dissolved oxygen or some acids are present in the mixture. The other, known as **colour quenching**, is absorption of some of the light emitted by the scintillator. Since many biological sample have some yellow, brown or green colour and the scintillator emission is blue, colour quenching is quite common.

Look back to Section 3.1 if you need to check why yellowish colours absorb blue light.

*Channels in scintillation counting*

One of the tasks in setting up a scintillation system is to set the ranges of energy that it will respond to. These are known as **channels** and are chosen to suit the particular isotope being counted. You can consult reference books or tables for values of their energies (Table 7.2 only gives a qualitative description).

Faires and Boswell's book (see 'Further reading' at the end of the chapter) gives a lot of detail about many radioactive isotopes.

You get different efficiencies of counting if you vary the channel settings and so you try to find the most efficient for your samples. The efficiency is determined by comparing the ratio of counts in one channel to those in another for standards of known activity. There is an empirical correlation between the ratio and the efficiency, and a correlation graph can be drawn manually or the data may be processed by a computer. Quenching also affects the counting efficiency and you cannot be sure that there is the same amount of quenching in each sample. However, the channels ratio is determined together with the channel count of the isotope of interest, so the efficiency can be found and the count rate converted to the activity of the sample.

Modern systems can count in two or more channels simultaneously so you can count two isotopes in the same sample. Thus, you would choose a range of fairly low energies for carbon-

14. Phosphorus-32 would give hardly any counts in this range and you would use a higher range. However, when the channels overlap, as they do for carbon-14 and tritium, you have to use some maths to obtain the two separate activities. In routine work, these days it is most likely that all these channel settings and mathematical procedures will be stored in a computer program and called up for each particular sample.

## 7.4 Analytical applications of radioactivity

### 7.4.1 Isotope dilution analysis

This is an analytical technique which makes use of measurements of specific activity and can be adapted to determine concentrations in test samples or to measure volumes that would be difficult to measure in any other way.

*Determination of concentrations*

You start with a radioactive standard of known specific activity; this must be the same chemical substance as your target for analysis. You add a measured portion of the standard to your test sample. The non-radioactive substance in the sample dilutes the added radioactivity. After thorough mixing, you isolate some of the substance and measure its reduced specific activity. The proportion by which it is reduced enables you to calculate the proportional dilution and hence the amount of non-radioactive substance in the sample.

For example, you might put 1 kBq of chloride-36 of specific activity $1\,\text{kBq}\,\text{mg}^{-1}$ into a sample of $100\,\text{cm}^3$ of urine and, after mixing, isolate the chloride as a known weight of silver chloride. If you then found that the specific activity of the chloride-36 was $0.01\,\text{kBq}\,\text{mg}^{-1}$ you would know that there was an overall 100-fold dilution. So there must have been 99 times as much non-radioactive chloride in the urine as was added in the standard. Since 1 kBq of the standard chloride-36 corresponded to 1 mg there must have been 99 mg of chloride in the $100\,\text{cm}^3$ of urine. This is a rather low level compared with normal urine and so it might be used in a medical diagnosis.

*Determination of hidden volumes*

Some volumes are difficult to measure, such as the blood volume of an experimental animal or the volume of water in underground water system. You can use the isotope dilution method and adapt the calculation so that the volume of the test system is found.

Such determinations do not have to use radioactivity; you just need to be able to measure the different concentrations to a sufficient degree of precision. Isotope dilution analysis was devised when other methods did not have the capability of very low-level detection. It is, today, often done more easily using other methods such as ion-selective electrodes (Section 4.4.1).

### 7.4.2 Tracers and metabolic studies

Tracers are substances encountered in biological work containing one or more radioactive atoms in their molecules. They are therefore chemically identical to the non-radioactive molecules and take part in all their metabolism. The result is that metabolic pathways can be followed as the radioactivity progresses through a variety of possible metabolites. The method uses thin-layer chromatography to a large extent with a Geiger counter that can move over the plate (a scanner) or autoradiography (see next paragraph) to register which components of the analysis are radioactive. The mechanism of photosynthesis was determined in this way. Some ingenious methods of exposing plants to radioactive carbon dioxide for a very brief time enable workers to show that it is incorporated into glyceraldehyde early in the sequence and that this is then further converted into four other sugars, including fructose, which can be converted into glucose. It is important in such studies that amounts of radioactivity are measured as well otherwise you may misinterpret the results. You may need to distinguish between a major and minor metabolic product.

### 7.4.3 Autoradiography

This technique is also called radioautography. Radioactive emissions affect photographic film in the same way as X-rays. For autoradiography you put a radioactive chromatogram, say, next to such a film for a time which might be from a few hours to several days, depending on the level of the activity on the chromatogram. At the end of the time you develop the film in the normal way and find dark spots indicating where the radioactivity is on the chromatogram. The film must be put as close to the radioactivity as possible to minimize any blurring of the image. In any case, higher-energy $\beta$-radiation and $\gamma$-radiation travel further through materials and tend to give greater blurring than soft $\beta$-radiation. The film must be rigorously shielded from any light as well as any background radiation, especially any $\gamma$-radiation. You can use the technique for locating DNA fragments in fingerprinting experiments and you can measure the intensities of the spots or bands using a densitometer to determine the relative amounts of activity. You can also examine autoradiographs of tissue sections

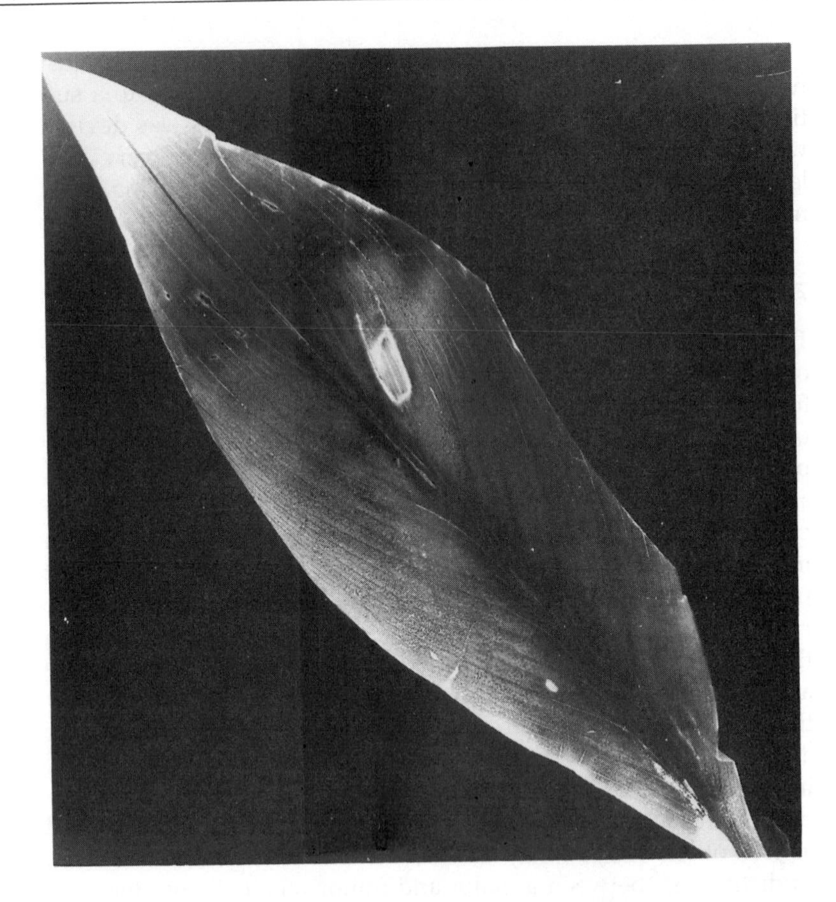

**Figure 7.5** Autoradiograph of leaf containing radioactive sucrose formed by photosynthesis in an atmosphere of $^{14}CO_2$.

to show the distribution of radioactivity at the cellular level. It is possible to examine such autoradiographs microscopically provided care has been taken to avoid blurring (loss of resolution).

*Pollution studies*

Small water-borne organisms such as *Daphnia* (water-fleas) accumulate heavy metals such as cadmium. You can use cadmium-109 to monitor the uptake and relate it to the concentration of cadmium in the water. Also, you can look for a correlation between the uptake and toxic effects of the metal.

*Radioimmunoassays*

This extremely sensitive technique has similarities to that of isotope dilution analysis but is based on the specific binding of antibodies to antigens. The proportions of radioactivity in bound and unbound fractions of the assay provide a means of determining the concentrations of substances which can be made to cause an immune response, such as proteins, hormones and drugs.

## Further reading

Chapman, J.M. and Ayrey, G. (1981) *The Use of Radioisotopes in the Life Sciences*, George, Allen & Unwin, ISBN 0 50 011799 2.
A brief but useful guide to the variety of biological uses of radioisotopes.

The ACOL 'teach yourself' style books include the following title.

Geary, W.J. (1986) *Radiochemical Methods*, Wiley, ISBN 0 471 91118 6.

For a detailed and comprehensive treatment of radioisotopes in all laboratory work see:

Faires, R.A. and Boswell, G.G.J. (1981) *Radioisotope Laboratory Techniques for Technicians*, 4th edn, Butterworths, ISBN 0 408 70940 5.

# This is not the end!

How did you get to this page? Are you the sort of person who turns to the end of a book to see whether you will like it? Or have you read through some, or all, of the chapters and feel you have finished with it? Whichever is the case, I would like to emphasize at this stage that you have really only begun but, from here on, you will probably not be able to study all of the techniques in detail. In fact, you could study any one of them for the rest of your career and still not exhaust the possibilities. You would almost certainly use the technique for new applications and you might well develop new methods based on the technique.

The more you use the techniques the more interesting they become. Some analysts give the impression that their technique is the only one worth pursuing. I hope you will not be like that! Of course, you will have a favourite technique and it will probably be the one most suited to your needs but you should always be alive to what other techniques can offer. On the other hand, you may feel that analysis is not for you. Nevertheless, I hope you have at least gained some idea of what analysis is and how it does it. If you continue in biological work, keep an eye open to see how analysis contributes to biological understanding.

Modern biological science, medical science, environmental science, forensic science, sports science, genetic research, quality control and many other disciplines use analytical techniques all the time and it is unlikely that further advances will be made without them. The power of analysis to unravel the complexities of biological molecules makes it one of the most interesting and challenging of all the scientific disciplines.

# Appendix 1 Some physical constants and some conversion factors

## Constants

The following constants are used in this book. They are quoted to three significant figures. Their meanings depend to some extent on the context in which they are used.

| | | |
|---|---|---|
| $R$ | $8.31\ \mathrm{J\,K^{-1}\,mol^{-1}}$ | Universal gas constant |
| $F$ | $96\,500\ \mathrm{C\,mol^{-1}}$ | Faraday constant |
| $c$ | $3.00 \times 10^{8}\ \mathrm{m\,s^{-1}}$ | Velocity of light in a vacuum |
| $h$ | $6.63 \times 10^{-34}\ \mathrm{J\,s}$ | Planck's constant |
| $N_a$ | $6.02 \times 10^{23}\ \mathrm{mol^{-1}}$ | Avogadro's constant |

## Conversion factors

These are needed when you come across older or non-SI units. You should convert to SI units before doing any calculations.

| | | | | |
|---|---|---|---|---|
| Wavelengths in spectroscopy | Å | Ångström unit | = | 0.1 nm |
| Radioactivity | Ci | curie | = | $3.7 \times 10^{10}$ Bq |
| Energy in radioactivity or spectroscopy | eV | electron-volt | = | $1.6 \times 10^{-19}$ J |

# Appendix 2
# The periodic table with values of relative atomic mass

| I | II | | | | | Transition Elements | | | | | | III | IV | V | VI | VII | Noble Gases |
|---|---|---|---|---|---|---|---|---|---|---|---|---|---|---|---|---|---|
| | | | | | | | | | | | | | | | | | 2 He HELIUM 4. 003 |
| 3 Li LITHIUM 6. 939 | 4 Be BERYLLIUM 9. 012 | | | | | | | | | | | 5 B BORON 10. 811 | 6 C CARBON 12. 011 | 7 N NITROGEN 14. 007 | 8 O OXYGEN 15. 999 | 9 F FLUORINE 18. 998 | 10 Ne NEON 20.183 |
| 11 Na SODIUM 22. 989 | 12 Mg MAGNESIUM 24. 312 | | | | | | | | | | | 13 Al ALLUMINIUM 26. 982 | 14 Si SILICON 28. 086 | 15 P PHOSPHORUS 30. 974 | 16 S SULPHUR 32. 064 | 17 Cl CHLORINE 35. 453 | 18 Ar ARGON 39. 948 |
| 19 K POTASSIUM 39. 102 | 20 Ca CALCIUM 40. 08 | 21 Sc SCANDIUM 44. 956 | 22 Ti TITANIUM 47. 90 | 23 V VANADIUM 50. 942 | 24 Cr CHROMIUM 51. 996 | 25 Mn MANGANESE 54. 938 | 26 Fe IRON 55. 847 | 27 Co COBALT 58. 933 | 28 Ni NICKEL 58. 71 | 29 Cu COPPER 63. 54 | 30 Zn ZINC 65. 37 | 31 Ga GALLIUM 69. 72 | 32 Ge GERMANIUM 72. 59 | 33 As ARSENIC 74. 922 | 34 Se SELENIUM 78. 96 | 35 Br BROMINE 79. 909 | 36 Kr KRYPTON 83. 80 |
| 37 Rb RUBIDIUM 85. 47 | 38 Sr STRONTIUM 87. 62 | 39 Y YTTRIUM 88 .905 | 40 Zr ZIRCONIUM 91. 22 | 41 Nb NIOBIUM 92. 906 | 42 Mo MOLYBDENUM 95. 94 | 43 Tc TECHNETIUM *99. 00* | 44 Ru RUTHENIUM 101. 07 | 45 Rh RHODIUM 102. 905 | 46 Pd PALLADIUM 106. 4 | 47 Ag SILVER 107. 870 | 48 Cd CADMIUM 112. 40 | 49 In INDIUM 114. 82 | 50 Sn TIN 118. 69 | 51 Sb ANTIMONY 121. 75 | 52 Te TELLURIUM 127. 60 | 53 I IODINE 126. 904 | 54 Xe XENON 131. 30 |
| 55 Cs CAESIUM 132. 905 | 56 Ba BARIUM 137. 34 | 57 La LANTHANUM 138. 91 | 72 Hf HAFNIUM 178. 49 | 73 Ta TANTALUM 180. 948 | 74 W TUNGSTEN 183. 35 | 75 Re RHENIUM 186. 2 | 76 Os OSMIUM 190. 2 | 77 Ir IRIDIUM 192. 2 | 78 Pt PLATINUM 195. 09 | 79 Au GOLD 196. 967 | 80 Hg MERCURY 200. 59 | 81 Tl THALLIUM 204. 37 | 82 Pb LEAD 207.19 | 83 Bi BISMUTH 208. 980 | 84 Po POLONIUM *209* | 85 At ASTATINE *210* | 86 Rn RADON *222* |
| 87 Fr FRANCIUM *223* | 88 Ra RADIUM *226* | 89 Ac ACTINIUM *227* | | | | | | | | | | | | | | | |

| LANTHANIDE SERIES | | | | | | | | | | | | | |
|---|---|---|---|---|---|---|---|---|---|---|---|---|---|
| 58 Ce CERIUM 140.12 | 59 Pr PRASEODYMIUM 140.907 | 60 Nd NEODYMIUM 144.24 | 61 Pm PROMETHIUM *147* | 62 Sm SAMARIUM 150.35 | 63 Eu EUROPIUM 151.96 | 64 Gd GADOLINIUM 157.25 | 65 Tb TERBIUM 158.924 | 66 Dy DYSPROSIUM 162.50 | 67 Ho HOLMIUM 164.93 | 68 Er ERBIUM 167.26 | 69 Tm THULIUM 168..934 | 70 Yb YTTERBIUM 173.04 | 71 Lu LUTETIUM 174.97 |
| **ACTINIDE SERIES** | | | | | | | | | | | | | |
| 90 Th THORIUM 232.038 | 91 Pa PROTACTINIUM *231* | 92 U URANIUM 238.03 | 93 Np NEPTUNIUM *237* | 94 Pu PLUTONIUM *244* | 95 Am AMERICIUM *243* | 96 Cm CURIUM *247* | 97 Bk BERKELIUM *247* | 98 Cf CALIFORNIUM *251* | 99 Es EINSTEINIUM *254* | 100 Fm FERMIUM *257* | 101 Md MENDELEVIUM *256* | 102 No NOBELIUM *256* | 103 Lr LAWRENCIUM *257* |

# Index

Page numbers appearing in **bold** refer to figures and page numbers appearing in *italic* refer to tables.